Nearly Zero Energy Building (NZEB) - Materials, Design and New Approaches

Edited by David Bienvenido-Huertas

Published in London, United Kingdom

Nearly Zero Energy Building (NZEB) - Materials, Design and New Approaches
http://dx.doi.org/10.5772/intechopen.97966
Edited by David Bienvenido-Huertas

Contributors
Avijit Ghosh, Shahaboddin Resalati, Claudia Naves David Amorim, Joara Cronemberger, Caio Frederico de Silva, Thiago Montenegro Góes, Ayana Dantas De Medeiros, João Manoel Dias Pimenta, Marco Antonio Egito, Adolfo Bauchspiess, Loana Nunes Velasco, José Manoel Morales Sánchez, Halil Zafer Alibaba, Jihui Yuan, Luis E. Juanicó, Nimish Biloria, Nastaran Abdollahzadeh, Karishma Kashyap, Usha Iyer-Raniga, Mary Myla Andamon, Suzana Domjan, Sašo Medved, Ciril Arkar, Rok Fink, Hüseyin Emre Ilgın, Markku Karjalainen, Marie Yli-Äyhö, Anu Soikkeli, Miguel Edgar Morales Udaeta, Fabio Pires, Fernanda Antonio, Claudia Terezinha de Andrade Oliveira, Lauri Metsäranta, Markku Norvasuo

First published in London, United Kingdom, 2022 by IntechOpen
IntechOpen is the global imprint of INTECHOPEN LIMITED, registered in England and Wales, registration number: 11086078, 5 Princes Gate Court, London, SW7 2QJ, United Kingdom

British Library Cataloguing-in-Publication Data
A catalogue record for this book is available from the British Library

Additional hard and PDF copies can be obtained from orders@intechopen.com

Nearly Zero Energy Building (NZEB) - Materials, Design and New Approaches
Edited by David Bienvenido-Huertas
p. cm.
Print ISBN 978-1-80355-312-2
Online ISBN 978-1-80355-313-9
eBook (PDF) ISBN 978-1-80355-314-6

For EU product safety concerns:
IN TECH d.o.o., Prolaz Marije Krucifikse Kozulić 3, 51000 Rijeka, Croatia,
info@intechopen.com or visit our website at intechopen.com.

Meet the editor

David Bienvenido-Huertas is an assistant professor in the Department of Building Construction, University of Granada, Spain. He is also a visiting professor at the University of La Coruña, Spain. His areas of expertise include climate change in the building sector, adaptive thermal comfort, heat transfer, fuel poverty, energy conservation measures, and the design of nearly zero-energy buildings. He is an author of more than fifty research papers and a recognized reviewer of various international indexed journals.

Contents

Preface **IX**

Section 1
Materials **1**

Chapter 1 **3**
Complementary Building Concept: Wooden Apartment Building: The Noppa toward Zero Energy Building Approach
by Markku Karjalainen, Hüseyin Emre Ilgın, Marie Yli-Äyhö and Anu Soikkeli

Chapter 2 **17**
An Aggregated Embodied and Operational Energy Approach
by Shahaboddin Resalati

Chapter 3 **35**
Coalash as Sustainable Material for Low Energy Building
by Avijit Ghosh

Section 2
Design **55**

Chapter 4 **57**
Wooden Facade Renovation and Additional Floor Construction for Suburban Development in Finland
by Markku Karjalainen, Hüseyin Emre Ilgın, Lauri Metsäranta and Markku Norvasuo

Chapter 5 **73**
An Integrated Design Process in Practice: A Nearly Zero Energy Building at the University of Brasília - Brazil
by Cláudia Naves David Amorim, Joara Cronemberger Ribeiro Silva, Caio Frederico e Silva, Thiago Montenegro Góes, Ayana Dantas de Medeiros, João Manoel Dias Pimenta, Marco Antonio Egito, Adolfo Bauchspiess, Loana Nunes Velasco and José Manoel Morales Sánchez

Chapter 6 99
Developing a Sustainable Solar-Residence Architecture Like a Home Unit without Energy Consumption from the Power Grid
by Fernanda Antonio, Claudia Terezinha de Andrade Oliveira, Fabio Pires and Miguel Edgar Morales Udaeta

Chapter 7 125
Evaluation of Energy Efficiency of Buildings Based on LCA and LCC Assessment: Method, Computer Tool, and Case Studies
by Suzana Domjan, Ciril Arkar, Rok Fink and Sašo Medved

Section 3
New Approaches 153

Chapter 8 155
Introduction of ZEB Technology in Japan
by Jihui Yuan

Chapter 9 173
Adaptive Thermal Comfort of an Office for Energy Consumption-Famagusta Case
by Halil Zafer Alibaba

Chapter 10 191
Holistic and Affordable Approach to Supporting the Sustainability of Family Houses in Cold Climates by Using Many Vacuum-Tube Solar Collectors and Small Water Tank to Provide the Sanitary Hot Water, Space Heating, Greenhouse, and Swimming Poole Heating Demands
by Luis E. Juanicó

Chapter 11 217
Highlighting the Design and Performance Gaps: Case Studies of University Buildings
by Karishma Kashyap, Usha Iyer-Raniga and Mary Myla Andamon

Chapter 12 235
Energy-Efficient Retrofit Measures to Achieve Nearly Zero Energy Buildings
by Nimish Biloria and Nastaran Abdollahzadeh

Preface

Currently, the construction industry has the most detrimental impact on the environment. On the one hand, constructions generate an environmental impact through the use of necessary resources and the generation of greenhouse gas emissions during the construction process. On the other hand, buildings consume large amounts of energy once they are in use. For this reason, it is necessary to develop efficient designs to mitigate the environmental impact of buildings during their different phases, with the goal of achieving a low-carbon building stock by 2050.

Energy policies are promoting decarbonization of the built environment using the nearly zero-energy building concept. However, the technological challenges to achieve nearly zero-energy buildings are great. As such, it is necessary to carry out research on innovative materials and designs. This book summarizes research in this area over three sections: "Materials," "Design," and "New Approaches." It includes twelve chapters that present current knowledge about nearly zero-energy buildings, which is crucial for academics, researchers, architects, engineers, and professionals in the building and construction sector.

David Bienvenido-Huertas
Department of Building Construction,
University of Granada,
Granada, Spain

Section 1

Materials

Chapter 1

Complementary Building Concept: Wooden Apartment Building: The Noppa toward Zero Energy Building Approach

Markku Karjalainen, Hüseyin Emre Ilgın, Marie Yli-Äyhö and Anu Soikkeli

Abstract

Increasing the construction of wooden apartment buildings has its place as part of preventing climate change. This chapter aims to explore the possibilities of expanding the construction of wooden apartment buildings on plots owned by the City of Helsinki in the Mellunkylä area by developing a series-produced wooden apartment building concept suitable for complementary construction—The Noppa concept. The sustainability of this approach is considered from the perspective of materials, construction methods, adaptability of the designed spaces, and housing design flexibility. In this study, the Noppa wooden apartment building concept with cross-laminated timber (CLT) elements has been developed varying in its facilities and architectural design features through architectural modeling programs to be used for complementary construction. The research findings are based on a theoretical approach that has not yet been practically tested but is proposed considering existing construction practices that need further investigation. It is believed that this chapter will contribute to the spread of wooden apartments to achieve a low-carbon economy as one of the key tools in tackling climate change problems. Particularly, proposed architectural design solutions will contribute to decarbonization of buildings as well as zero energy building (nZEB) approach.

Keywords: apartment building, zero energy buildings (nZEB), architectural design, timber/wood, CLT, Finland

1. Introduction

"Net Zero Energy Buildings" will be the next big frontier for innovation and competition in the world's real estate market and can be promptly scaled in Europe as in North America [1]. In this sense, European energy policies introduced the net zero energy building (nZEB) target [2] to promote the energy transition of the construction sector. EU programs, especially "Horizon 2020," introduce the nZEB design as well as its evolution to positive energy building (PEB) model [3]. Especially the construction industry is one of the main reasons for this problem due to excessive emissions to the environment [4] resulting from the processes of buildings' heating and cooling systems.

Until recently, Finnish building codes were only an incentive to construct low-energy buildings, and Finland had no legislation or guidelines on life-cycle emissions. However, like other Scandinavian countries working toward regional carbon neutrality, Finland targets carbon neutrality by 2035 and is developing policies, including low-carbon construction legislation [5]. Additionally, the Finnish Ministry of Environment has set a target for building life-cycle legislation to account for CO_2 emissions by 2025 [6]. The aim is to influence the total carbon footprint of the construction and the building heating carbon footprint of the energy used through financial incentives [7, 8]. The Finnish Ministry of Environment is considering financial controls over the life cycle of the building to reduce CO_2 emissions, 50 years building life is a set of target control plans [9, 10]. Like Finland's national goal, the Helsinki-Uusimaa Region aims for climate neutrality by 2035 [11].

In this sense, bio-based materials such as wood come to the fore with many advantages such as good indoor air quality, thermal insulation [12]. Bio-based materials are generally hygroscopic; that is, they retain water molecules until an equilibrium state of water content is reached for the relative humidity of the ambient air [13], which positively affects indoor air quality. The performance of these materials can significantly contribute to microclimate comfort by managing energy and mass (vapor) transfer. Furthermore, wood acts as a thermal insulator while also providing a suitable internal surface temperature. Timber also protects from thermal bridges, as it is one of the very few available materials capable of both load bearing and insulation. Wood's volumetric change due to heat is minimal; therefore, for example, in solid wood structures, in glued arrangements, it is considered a good structural material in many cases [14].

Wood construction stands out as one of our best allies in solving the climate crisis, thanks to its positive environmental characteristics such as low carbon emissions during processing and significant carbon storage in use. Additionally, according to life cycle assessment–based research in the literature [15–17] the selection of wood-based materials has a substantially lower impact on CO_2 emissions in comparison with non-wood-based materials as in the study on the life-cycle assessment of a wooden single-family house in Sweden [18]. Wood construction also supports the Finnish government's bio-economic strategy for a carbon-neutral society by 2035 and addresses European climate policy [19]. In particular, engineered wood products (EWPs) such as cross-laminated timber (CLT) are being used in increasingly demanding applications [20] to meet the sustainable construction challenge [21–23]. The many advantages of CLT include low carbon and high thermal insulation, excellent in-plane and out-of-plane strength, high strength-to-weight ratio, and large-scale and high-rise buildings to be built [24, 25].

On the other hand, Finnish residents generally welcome timber construction and multistory timber apartment buildings [26]. They attributed the features of timber apartment building residence such as good sound insulation, good indoor climate, beauty, warm atmosphere, and coziness. Furthermore, they wish for more wood as a visible surface material inside the building and more timber apartment buildings.

Thus, wood-based solutions have traditionally held a strong position in Finland's construction industry, with wood accounting for 40% of all building materials, and about 80% of single-family homes are timber-framed. About 12 million cubic meters of sawn wood were produced in Finland in 2018, and about four-fifths of the sawn wood was used for construction. Moreover, the National Wood Building Programme (2016–2022) in Finland aims to increase wood use and long-term carbon storage in wood structures by promoting the growth of internationally competitive industrial wood building knowledge and production [27].

The Noppa concept will be implemented in the New Housing Forms—Integration of Living Suburbs (AsuMut) project in collaboration with Tampere University and the City of Helsinki. This project is part of a suburban program managed and funded by the Finnish Ministry of Environment. Three of the cities in the Helsinki suburban program relate to the urban reform area, Malminkartano-Kannelmäki, Malmi, and Mellunkylä. The Helsinki suburban program is connected in addition to several strategic programs of the City of Helsinki, such as the Helsinki City Strategy for 2017–2022, Carbon neutral Helsinki 2035 action program, and Implementation program for housing and related land use [28]. The City of Helsinki aims for carbon neutrality by 2035 and uses wood instead of the concrete structure to achieve this. Changing the segregation of existing residential areas to strengthen their attractiveness creates prosperity for the present and future residents of the area. By increasing the construction of wooden apartment buildings in complementary constructions, the naturalness of wood can bring comfort and humanity to the suburbs.

The focus of the study is the wooden structure development of an apartment concept, where complementary construction projects of mass-produced wooden apartments can be designed. Using the concept, it is possible to design wooden buildings in Helsinki and others in the growth centers of our country with an architectural environment that differs in building stock and additional site requirements. This study targets Mellunkylä, one of the Helsinki-owned plots where the possibility of complementary construction is being considered.

In this context, architectural design has an important opportunity to support sustainable development [29]. In Finland, this will also be promoted toward the end of 2020, graduating from the architectural policy program proposal of the Ministry of Education and Culture as well as the Ministry of the Environment [30], with the main theme being combating climate change and sustainability toward sustainable architecture. In this sense, architects can make a great contribution to a constructive building culture by ensuring the ecological quality and sustainability of the living environment.

On the other hand, it is worth mentioning here that as the population concentrates in cities and available land is depleted, housing flexibility is becoming an essential feature in the transformations of our daily lives [31]. Housing flexibility, which is associated with different typologies, provides the opportunity to change buildings spatially or structurally to meet the needs of building occupants by adapting to technological, cultural, and economic changes that have occurred over time [32]. Housing resilience is based on sustainable consumption in line with building life extension, recycling, and waste management [33]. Today, the need for flexibility in the housing field has become very urgent, which is a fundamental feature of architecture [34]. In this study, housing flexibility is also considered an important architectural design input and contributes to the nZEB approach in terms of its resilient features such as recycling.

Overall, this chapter aims to create higher value-added circular economy opportunities to promote the competitiveness of large-scale industrial timber construction at the local level and to support European climate policy as part of a low-carbon economy. It is believed that this study will help the dissemination of wooden apartment buildings for different and innovative architectural applications as one of the key tools to contribute to decarbonization of buildings and nZEB approach.

2. Research method

This study was carried out with architectural modeling methods used in the solution of research and design problems in architectural activities. This method enables architects to think, write, discuss, and disseminate as a bridge from theory to practice [35]. It is widely used in architectural design research where architects use it as a tool for research methodology [36, 37].

Additionally, at present, there is no single approach to making the object and subject of architectural activity, which inevitably leads to significant differences in research methods and architectural design of objects, especially at such important levels of solving this problem [38]. On the other hand, the precise operation of text and project interaction in architectural design research remains a highly debated and relatively unformed topic [39–42].

Therefore, in this study, main business applications such as AutoCAD, SketchUp, parametric modeling and information modeling methodology of buildings, and complex object modeling methods used in modern architectural design applications (e.g., [43, 44]) were employed. Here, creative proposals are realized through a mix of drawings and models as visual representations to encourage a fresh and lively approach to architectural research. **Figure 1** shows the architectural design steps used in this study as the research method with numerous background variables (e.g., client/user needs and aspirations, project philosophy, design idea and inspiration, marketing, project management, material research, operation management).

Starting points were prepared for the Mellunkylä region (**Figure 2**) to establish the design principles, which have been approved by the City of Helsinki's Urban Environment Board as a basis for further planning in September 2020. The aim of

Figure 1.
Architectural design steps used in this study as the research method.

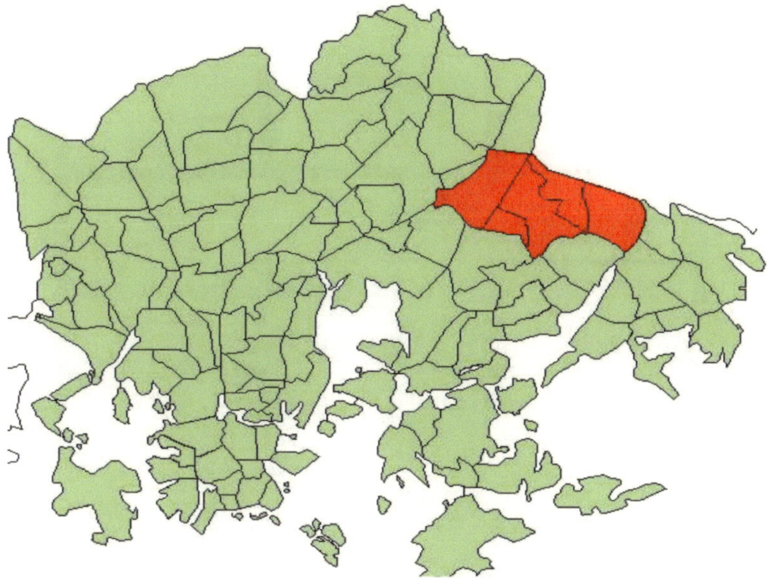

Figure 2.
Mellunkylä region as a district of Helsinki.

the urban reform is to increase the attractiveness of the region by boosting housing and employment, improving accessibility, and enhancing the district public service network together with reducing CO_2 emissions and contributing to nearly zero-energy buildings. According to the design principles in complementary constructions, the aim is to preserve the typical features of the site, as well as their natural environment.

The objective of Noppa approach, the solid house frame apartment concept to be produced in series, is to be a step toward a smoother wooden apartment construction. The starting point of the Noppa concept is to provide functional, aesthetic, and affordable housing facilities with efficient wooden design solutions for different construction site conditions.

As its construction principles, the Noppa approach has a narrow frame and is suitable for its size for well-finished construction sites. If there is space beyond the additional site building, the Noppa apartment building can be converted into an apartment building with two or multiple stairs connecting the short side of the house. The Noppa apartment has a clear basic framework, and its facilities and layout are highly adaptable, where efforts have been made to select feasible structural solutions that are as simple as possible. Standardized structural solutions allow different collaboration of actors in construction chips, and the construction concept can be developed and implemented by several interested parties. The building plan meets the requirements of current legislative building codes in Finland such as the Finnish fire code.

In apartments' floors, volume elements placed on the base layer have 10 apartments of varying sizes (from 53 to 134 m^2) and types. The living areas of the apartments are of reasonable size, and the smallest residences allowed in the regulations cannot be found in the selection of 20 m^2. Adequate sizing of dwellings increases living comfort, the ability of the building to adapt to changing housing needs, and thus longevity. The goal in the design is the premises of the apartment flexible space solutions for functional use. The plan focuses on enabling a diverse mix of housing and transformative spaces within the residences, thereby increasing the value of the

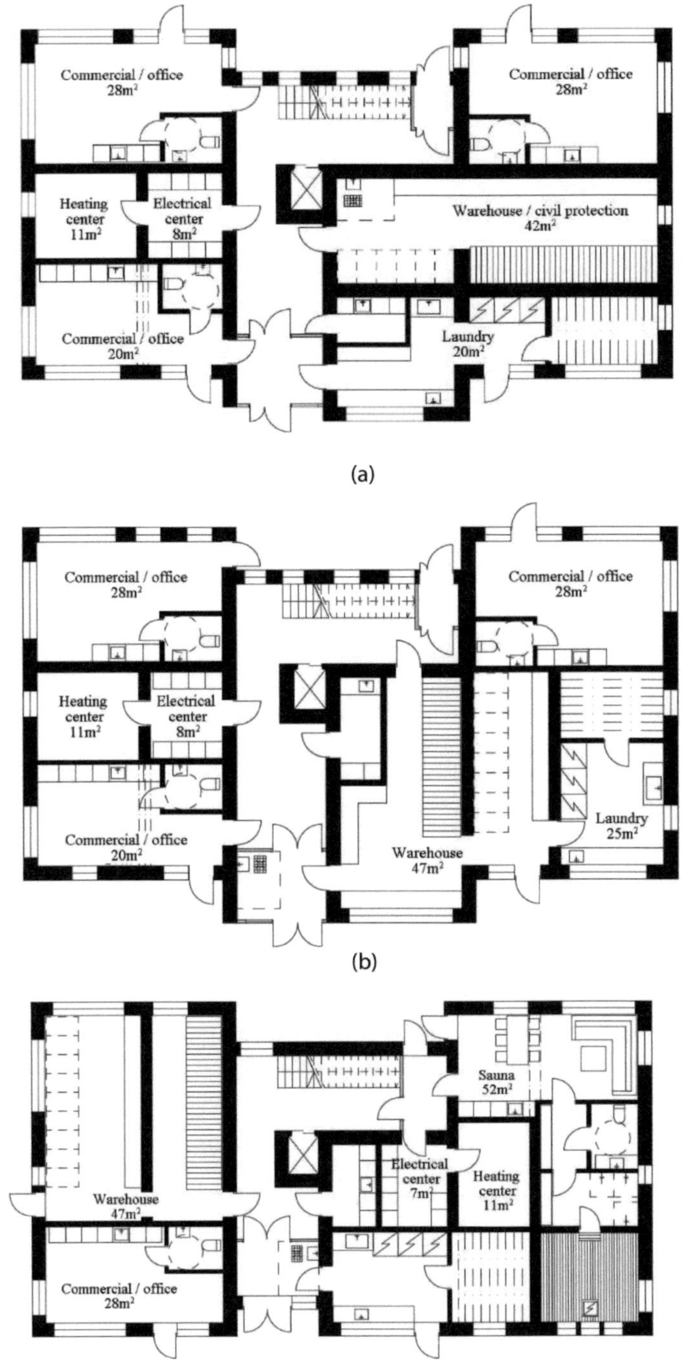

Figure 3.
Ground floor alternatives: (a) with warehouse/civil protection; (b) no shelter; and (c) with sauna and no shelter.

building in the long run. For ground floors, three different options (**Figure 3**) are provided with all necessary technical services, while many different types of apartments' living floors (**Figure 4**) are proposed.

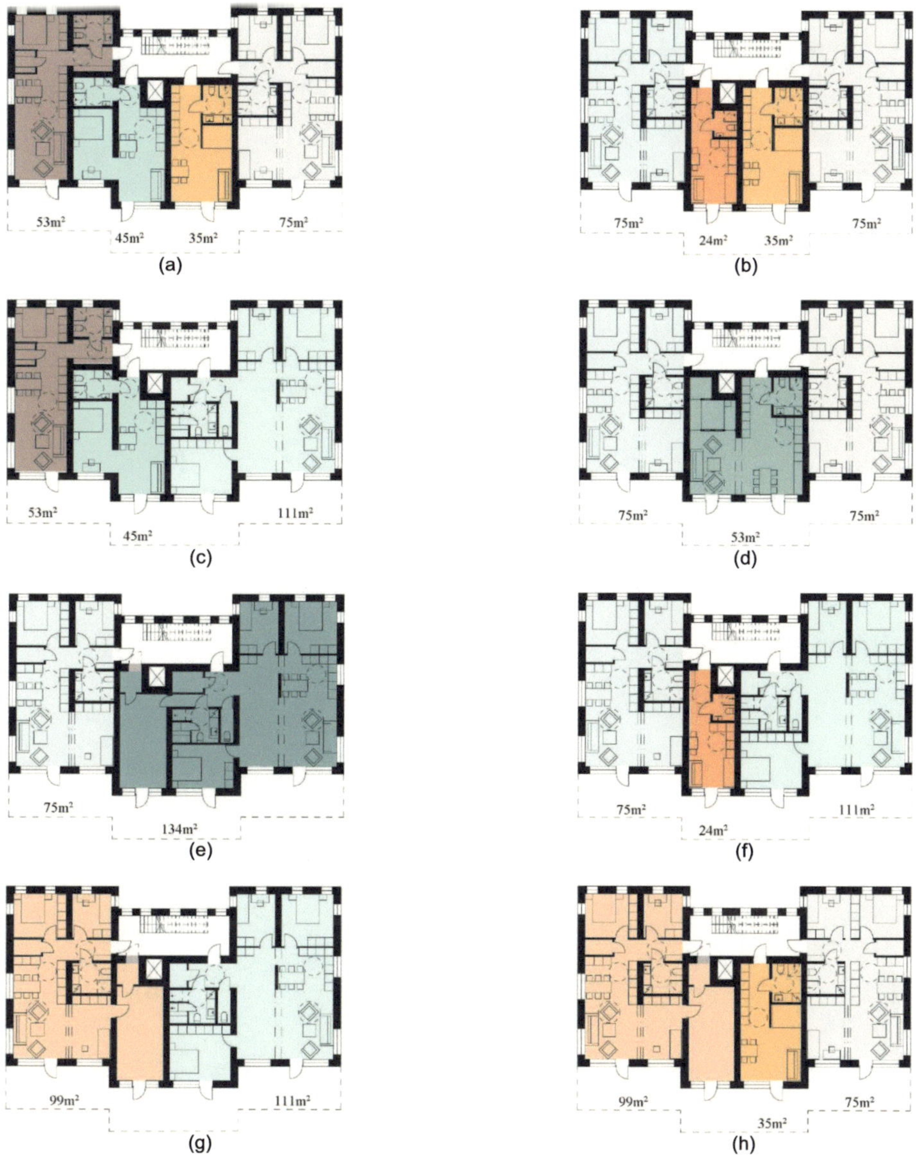

Figure 4.
Living floor alternatives (a–h).

The choice of gable roof supports practical functionality in water management, which is essential for the longevity of a timber-framed apartment building (**Figure 5**). Besides natural ventilation, the use of gravity as a basic solution also requires the shape of the roof to create the height difference necessary for gravity ventilation to be created. The Noppa basic solution has four gable roof options that affect the architecture of the building, for example, the upper tiers space arrangements. There are three balcony solutions, a flat and shaped balcony area across the entire facade, and freestanding, self-contained balcony towers (**Figure 6**).

Figure 7 shows 3D views and typical floor plans of four different alternatives for the Noppa basic solution.

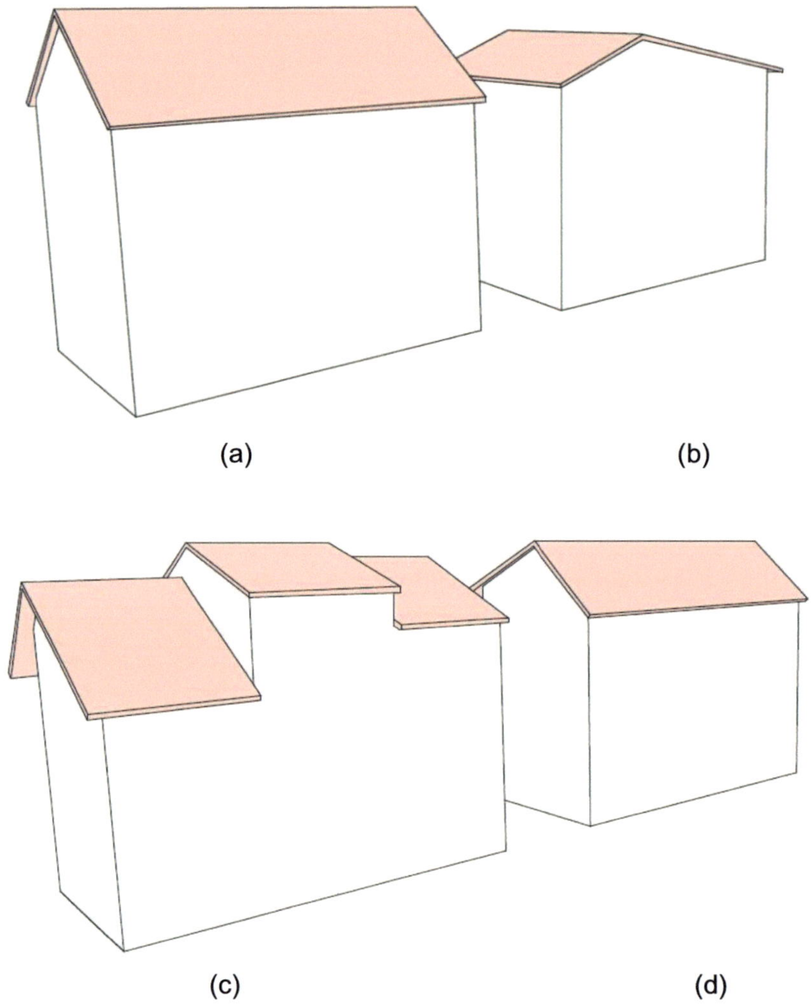

Figure 5.
Gable roof alternatives: (a) symmetrical gable roof; (b) inverted gable roof; (c) partitioned gable roof; and (d) asymmetrical gable roof.

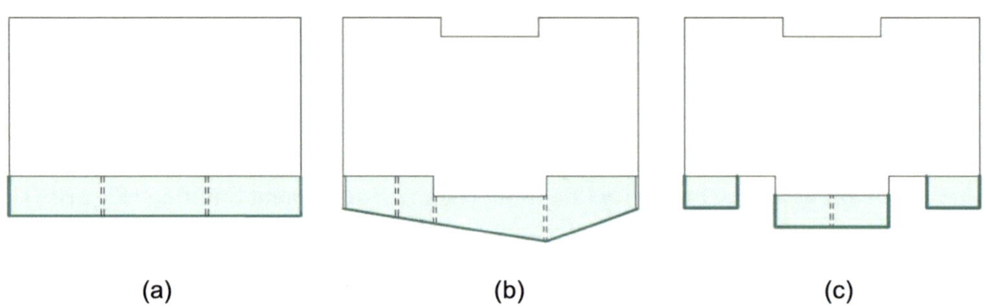

Figure 6.
Balcony alternatives: (a) full facade balcony; (b) shaped balcony; and (c) individual balconies.

(a)

(b)

Figure 7.
Different design alternatives for the Noppa basic solution (a–d).

3. Conclusion

This chapter aimed to search for the possibilities of expanding the construction of wooden apartment buildings in the Mellunkylä region by developing a mass-produced wooden apartment concept suitable for complementary construction—"The Noppa concept." The sustainability of this concept was considered from the perspective of materials, construction methods, the adaptability of the designed spaces as well as design flexibility. The results were the architectural design proposals based on a theoretical approach considering contemporary applications in the wooden apartment construction market, but further research such as life-cycle assessment will be done as part of other studies.

As a country with a sustainable social structure, a well-educated population, and a high level of technological expertise, Finland has an excellent opportunity to rebuild itself in line with the principles of sustainable development and zero energy building as in the case of Mellunkylä region. Advances in research and product development related to (engineered) wood products with high processing value and long carbon storage times, sustainable use of industry side streams, and ensuring transparency and efficiency in the timber market will contribute to this sustainable development. Furthermore, encouragement of wood structures to function as carbon storage, endorsing material neutrality in fire regulations to reduce the need for double fire protection of wood buildings, and industries and other private investors' contributions to sustainable development by focusing on improving existing processing technologies and making them more resource and energy-efficient play a critical role in this progress.

In this sense, it is believed that this chapter will contribute to the spread of wooden apartments to achieve a low-carbon economy as one of the key tools in tackling climate change problems. In particular, the proposed architectural design solutions will support the decarbonization of buildings and a zero-energy building approach.

Funding

Based on Marie Yli-Äyhö's MSc thesis (Tampere University), this project was funded by Finland's Ministry of Environment and ARA (The Housing Finance and Development Centre of Finland) as a part of the AsuMut (New forms of housing—Unifying neighborhoods) project.

Author details

Markku Karjalainen[1], Hüseyin Emre Ilgın[1*], Marie Yli-Äyhö[1] and Anu Soikkeli[2]

1 Tampere University, Tampere, Finland

2 University of Oulu, Oulu, Finland

*Address all correspondence to: emre.ilgin@tuni.fi

References

[1] Shady A. Net Zero Energy Buildings (nZEB)—Concepts, Frameworks and Roadmap for Project Analysis and Implementation. Amsterdam, Netherlands: Elsevier; 2018

[2] European Commission. Nearly Zero-Energy Buildings. 2021. Available from: https://ec.europa.eu/energy/topics/energy-efficiency/energy-efficient-buildings/nearly-zero-energy-buildings_en [Accessed: October 25, 2021]

[3] BPIE (Buildings Performance Institute Europe). Positive Energy Buildings. 2014. Available from: https://www.bpie.eu/event/positive-energy-buildings-wishful-thinking-or-built-reality/ [Accessed: October 25, 2021]

[4] Magrini A, Lentini G, Cuman S, Bodrato A, Marenco L. From nearly zero energy buildings (nZEB) to positive energy buildings (PEB): The next challenge—The most recent European trends with some notes on the energy analysis of a forerunner PEB example. Developments in the Built Environment. 2020;**3**:100019. DOI: 10.1016/j.dibe.2020.100019

[5] Kuittinen M, Häkkinen T. Reduced carbon footprints of buildings: New Finnish standards and assessments. Buildings and Cities. 2020;**1**(1):182-197. DOI: 10.5334/bc.30

[6] Kivimaa P, Kangas H, Lazarevic D, Lukkarinen J, Åkerman M, Halonen M, Nieminen M. Transition Towards Zero Energy Buildings Insights on Emerging Business Ecosystems, New Business Models and Energy Efficiency Policy in Finland, Finnish Environment Institute. 2019. Available from: https://helda.helsinki.fi/bitstream/handle/10138/293607/SYKEju_5_Transition-towards-zero-energy-buildings.pdf?sequence=1 [Accessed: October 25, 2021]

[7] Salo M, Nissinen A. Finnish Environment Institute Consumption choices to decrease personal carbon footprints of Finns, Reports of The Finnish Environment Institute 30. 2017. Available from: https://media.sitra.fi/2017/10/23144245/Consumption_choices_to_decrease_personal_carbon_footprints_of_Finns.pdf [Accessed: October 25, 2021]

[8] Finland has an excellent opportunity to rebuild itself in line with the principles of sustainable development, Carbon neutral Finland, Finnish Government. 2021. Available from: https://valtioneuvosto.fi/en/marin/government-programme/carbon-neutral-finland-that-protects-biodiversity [Accessed: October 25, 2021]

[9] Reform of the Climate Change Act. Finnish Ministry of Environment. 2021. Available from: https://ym.fi/en/the-reform-of-the-climate-change-act [Accessed: October 25, 2021]

[10] Integrated Reporting on Greenhouse Gas Policies and Measures and on Projections under Article 18 of Regulation (EU) No 2018/1999 and Articles 36, 37 and 39 of Commission Implementing Regulation (EU) No 2020/1208, VN/4694/2021. European Commission. 2021. Available from: https://tem.fi/documents/1410877/2132096/Integrated+reporting+on+PAMs_Finland_2021.pdf/26c216d7-85ef-d7b5-ce99-67e1283fa0f0/Integrated+reporting+on+PAMs_Finland_2021.pdf?t=1616157896988 [Accessed: October 25, 2021]

[11] Helsinki-Uusimaa Regional Council. Climate Neutral Helsinki-Uusimaa 2035.

2021. Available from: https://www.uudenmaanliitto.fi/en/development_and_planning/regional_programming/climate_neutral_helsinki-uusimaa_2035 [Accessed: October 25, 2021]

[12] Sandak A, Sandak J, Brzezicki M, Kutnar A. Designing building skins with biomaterials. In: Bio-Based Building Skin. Environmental Footprints and Eco-design of Products and Processes. Singapore: Springer; 2019. DOI: 10.1007/978-981-13-3747-5_3

[13] ASHRAE Standard. Thermal Environmental Conditions for Human Occupancy. Atlanta: American National Standards Institute; 2004

[14] Herzog T, Natterer J, Schweitzer R, Volz M, Winter W. Timber Construction Manual. Basel: Birkhäuser; 2012

[15] Liang S, Gu H, Bergman R, Stephen SK. Comparative life-cycle assessment of a mass timber building and concrete alternative. Wood and Fiber Science. 2020;**52**(2):217-229

[16] Anwar S. Life Cycle Analysis and Comparison of Different Construction Materials of Residential Buildings for Sustainable Construction Choice, HTW Berlin—(Hochschule für Technik und Wirtschaft Berlin). Berlin, Germany: University of Applied Sciences; 2020

[17] Liang S, Gu H, Bilek T, Bergman R. Life-Cycle Cost Analysis of A Mass-Timber Building: Methodology and Hypothetical Case Study. Madison, WI: U.S. Department of Agriculture, Forest Service, Forest Products Laboratory; 2019. pp. 1-11

[18] Petrovic B, Myhren JA, Zhang X, Wallhagen M, Eriksson O. Life cycle assessment of a wooden single-family house in Sweden. Applied Energy. 2019;**251**:113253. DOI: 10.1016/j.apenergy.2019.05.056

[19] Wood Building Programme. Finnish Ministry of Environment. 2020. Available from: https://ym.fi/en/wood-building [Accessed: October 25, 2021]

[20] Stepinac M, Šušteršic I, Gavric I, Rajcic V. Seismic design of timber buildings: Highlighted challenges and future trends. Applied Sciences. 2020;**10**:1380

[21] Brandner R, Flatscher G, Ringhofer A, Schickhofer G, Thiel A. Cross laminated timber (CLT): Overview and development. European Journal of Wood and Wood Products. 2016;**74**:331-351. DOI: 10.1007/s00107-015-0999-5

[22] Gasparri E, Lucchini A, Mantegazza G, Mazzucchelli ES. Construction management for tall CLT buildings: From partial to total prefabrication of façade elements. Wood Material Science and Engineering. 2015;**10**(3):256-275. DOI: 10.1080/17480272.2015.1075589

[23] Lukacs I, Björnfot A, Tomasi R. Strength and stiffness of cross-laminated timber (CLT) shear walls: State-of-the-art of analytical approaches. Engineering Structures. 2019;**178**:136-147

[24] Shahnewaz M, Tannert T, Alam MS, Popovski M. In-plane stiffness of cross-laminated timber panels with openings. Structural Engineering International. 2017;**27**:217-223

[25] Sikora KS, McPolin DO, Harte AM. Effects of the thickness of cross-laminated timber (CLT) panels made from Irish Sitka spruce on mechanical performance in bending and shear. Construction and Building Materials. 2016;**116**:141-150

[26] Karjalainen M, Ilgın HE. The change over time in finnish residents' attitudes towards multi-story timber apartment buildings. Sustainability. 2021;**13**:5501. DOI: 10.3390/su13105501

[27] The Finnish Ministry of Agriculture and Forestry. Wood Construction is Being Promoted in Finland. Available from: https://mmm.fi/en/en/forests/use-of-wood/wood-construction [Accessed: October 25, 2021]

[28] The City of Helsinki. Suburban Regeneration. 2021. Available from: https://www.uuttahelsinkia.fi/en/sustainable-urban-development/suburban-regeneration/suburb-programme [Accessed: October 25, 2021]

[29] Feria M, Amado M. Architectural design: Sustainability in the decision-making process. Buildings. 2019;**9**:135. DOI: 10.3390/buildings9050135

[30] Ministry of Education and Culture. Apoli2020: New Architectural Policy Programme. 2020. Available from: https://minedu.fi/en/apoli2020 [Accessed: October 25, 2021]

[31] Paris SRD, Lopes CNL. Housing flexibility problem: Review of recent limitations and solutions. Frontiers of Architectural Research. 2018;**7**(1):80-91. DOI: 10.1016/j.foar.2017.11.004

[32] Canepa S. Living in a flexible space. IOP Conference Series: Materials Science and Engineering. 2017;**245**. DOI: 10.1088/1757-899X/245/5/052006

[33] Gilani G, Türker ÖO. Assessing flexibility in real estate mass housing. Arquitetura Revista. 2020;**16**(1):154-175. DOI: 10.4013/arq.2020.161.09

[34] Magdziak M. Flexibility and adaptability of the living space to the changing needs of residents. IOP Conference Series: Materials Science and Engineering. 2019;**471**:072011. DOI: 10.1088/1757-899X/471/7/072011

[35] Fraser M. Design Research in Architecture: An Overview. London: Ashgate; 2013

[36] Akšamija A. Research Methods for the Architectural Profession. New York, NY: Routledge; 2021

[37] Short CA. What is 'architectural design research'? Building Research and Information. 2008;**36**(2):195-199. DOI: 10.1080/09613210701811015

[38] Vasilenko NA. General system principles of architectural systems formation. IOP Conference Series: Materials Science and Engineering. 2020;**753**

[39] Luck R. Design research, architectural research, architectural design research: An argument on disciplinarity and identity. Design Studies. 2019;**65**:152-166. DOI: 10.1016/j.destud.2019.11.001

[40] Gao Y. Design research in architecture: An overview edited by Murray Fraser. The Design Journal. 2015;**18**(2):291-294. DOI: 10.2752/175630615X14212498964439

[41] Eilouti B. Scenario-based design: New applications in metamorphic architecture. Frontiers of Architectural Research. 2018;**7**(4):530-543. DOI: 10.1016/j.foar.2018.07.003

[42] Luck R. Participatory design in architectural practice: Changing practices in future making in uncertain times. Design Studies. 2018;**59**:139-157. DOI: 10.1016/j.destud.2018.10.003

[43] Bhooshan S. Parametric design thinking: A case-study of practice-embedded architectural research. Design Studies. 2017;**52**:115-143. DOI: 10.1016/j.destud.2017.05.003

[44] Fu F. Design and Analysis of Tall and Complex Structures. Oxford and Cambridge: Butterworth-Heinemann, Elsevier; 2018

Chapter 2

An Aggregated Embodied and Operational Energy Approach

Shahaboddin Resalati

Abstract

Highly insulated envelopes are an integral part of any net zero energy building with a target to reduce the demand that need to be supplied by the renewable energy and other mitigating measures. While stricter insulation levels can in theory reduce the operational energy demand of buildings, the additional embodied energy investment in the insulations can become significant and not recovered within the expected timeframes. Accounting for embodied energy investment requires a paradigm shift in design of highly insulated buildings and can determine U-value levels that can be justified based on an aggregated operational and embodied energy approach. The following chapter discusses the aggregated approach in more detail showcasing the shortcomings of existing building codes and standards using a case study building. The chapter also reviews the potential barriers of adopting such approaches with a specific focus on the uncertainties of embodied energy data and offers a holistic view on its implications for various end-users and stakeholders within the construction sector. The presented analyses in this chapter depict optimal insulation levels beyond which the additional embodied energy burden cannot be recovered using the associated operational energy savings highlighting the necessity of accounting for embodied energy in developing future design principles for zero energy buildings.

Keywords: embodied carbon, aggregated carbon, insulation materials, optimum carbon levels, Life Cycle Assessment

1. Introduction

The international building codes and standards have consistently, through their various iterations, sought to reduce the energy demand of buildings with a focus on better fabric performance and lower U-value requirement among others (**Table 1**). This was simply due to the fact that the operational energy demand, in earlier versions of standards, was taken as 10 times greater than the embodied energy load, and therefore reasonable to be given priority [2–6].

More recently however, when reducing the carbon emissions from the built environment came under more serious scrutiny, this trend that has been cemented in building standard around the world has been questioned and analysed further and different countries have started acknowledging embodied energy in their regulations. For example, France and Belgium are pioneering the move to mandate consideration of embodied energy in their building regulations in Europe. Although this is still relatively new

		1985	2003	2007	2010	2012	2018
Finland	U-value wall	0.28	0.25	0.24	0.17	0.17	0.17
	U-value roof	0.22	0.16	0.15	0.09	0.09	0.09
		2006		2010		2013	
		Notional building	Limiting factor	Notional building	Limiting factor	Notional building	Limiting factor
UK	U-value wall	0.35	0.7	0.2	0.3	0.18	0.3
	U-value roof	0.25	0.35	0.13	0.2	0.13	0.2
		EnEV 2002	EnEV 2004	EnEV 2007	EnEV 2009	EnEV 2014	EnEV 2016
Germany	U-value wall	0.45	0.45	0.45	0.35	0.35	0.28
	U-value roof	0.45	0.45	0.45	0.35	0.35	0.28
		BFS 2008		BFS 2011		BFS 2016	
Sweden	U-value wall	0.18		0.18		0.18	
	U-value roof	0.13		0.13		0.13	
		BCA2007		NCC 2011		NCC 2015	
Australia	U-value wall	0.52 (Z1, 2 and 3)–0.3 (zone 8)		0.35 (Z1–Z7)–0.26 (Z8)		0.35 (Z1–Z7)–0.26 (Z8)	
	U-value roof	0.37 (Z1)–0.21 (Z8)		0.2 (Z1–Z7)–0.16 (Z8)		0.2 (Z1–Z7)–0.16 (Z8)	
		IECC 2009		IECC 2012		IECC 2018	
United States	U-value wall	0.197–0.057 (Z1–Z8)		0.197–0.057 (Z1–Z8)		0.197–0.057 (Z1–Z8)	
	U-value roof	0.035–0.026 (Z1–Z8)		0.035–0.026 (Z1–Z8)		0.035–0.026 (Z1–Z8)	

Adapted from Resalati et al. [1].

Table 1.
Changes in building codes and standards around the world.

and low impact and the building product manufacturers are only required to report Life Cycle Assessment (LCA) data should they decide to promote the environmental performance of their products, it is a significant shift towards regulating embodied energy in buildings [7]. Other countries within Europe joining the initiative include Austrian, the Netherlands and German legislations. These although acknowledge embodied energy investment a significant contributor to the overall carbon footprint of new buildings, only focus on operational energy currently. Although not fully incorporated in building regulations there exist examples of various embodied energy inventories dedicated to the construction sector and the associated materials and products including BRE's Green Guide and the Inventory of Carbon and Energy (ICE), U.S. Life Cycle Inventory Database, and the Canadian Building Material Life Cycle Inventory Database [1].

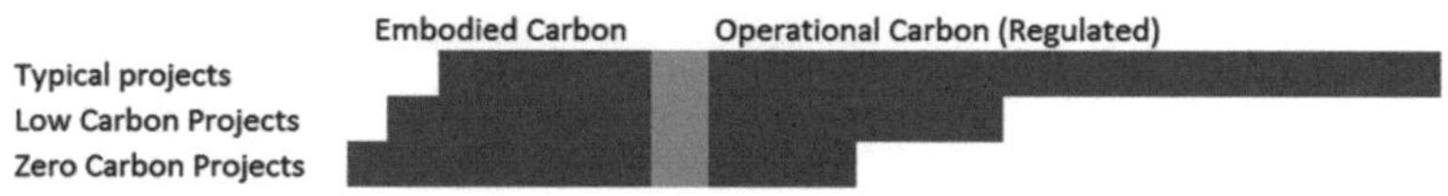

Figure 1.
Embodied to operational proportions for low and zero carbon buildings.

In recent years, the increased use of LCA evaluations to measure the environmental performance of building materials and products has emerged from the push toward integrating embodied energy in emission equations. Various environmental certification systems have been developed and used, such as the Environmental Product Declaration (EPD) [8], to independently verify documents that transparently and accurately communicate the environmental impact of various products in accordance with EN 15804 and ISO 14025. EPDs are type III environmental declarations based on the fundamental product category rules of European standards (PCR).

Although embodied energy has been acknowledged in regulations and researched substantially in the literature, it is still not fully regulated. The ratio of embodied to operational energy has changed over the years with the operational energy reducing as a result of increased adoption of renewable energy and better fabric standards. This, at the same time, increased the use of insulation in the buildings and shifted the ratio considerably [9]. As the ratio shifts, future low and zero energy buildings may see comparable embodied and operational energy measures, or even embodied energy outweighing the operational energy (**Figure 1**), concluded in RICS [10], Kristjansdottir et al. [11], Sartori & Hestnes [12], Dixit [13], Chau et al. [14], Stephan et al. [15], Dascalaki et al. [16], Mourao et al. [17], Gustavsson & Joelsson [18], Azari and Abbasabadi [19], and Dascalaki et al. [20]. Such drastic changes necessitate a thorough examination of the constraints and challenges that come with regulating embodied energy in the construction industry.

The following section examines the relative challenges and arising opportunities, focusing on issues such as the consistency and reliability of existing data data utilised in LCA analysis, as well as inconsistent modelling methodologies that produce outputs with a high level of uncertainty, and lays the foundation for future research.

2. Embodied energy data: challenges and opportunities

LCA studies are used to properly assess a product's or service's environmental impact. This strategy is largely data-driven, and is heavily reliant on the availability of precise, dependable, and high-quality data [21]. Gathering data with such qualities, on the other hand, has proven difficult for LCA end users and practitioners [22] due to a variety of issues, including manufacturer confidentiality requirements, the time and expertise required to generate reliable data, and inconsistent application of methodological approaches to data analysis [23, 24].

In theory, any LCA study in accordance with common PCRs should allow for reliable comparative analyses to take place for various building materials, products, and services. In practice however, the assumptions used in the LCA models including the service life of the product, maintenance requirements, in use operating energy, and varying system boundary options [1, 25] can significantly shift the results of the LCA models [26]. Several researchers have reported disparities between LCA results based on fundamentally different assumptions, such as functional units [27], system boundaries [28, 29],

LCI databases [21, 30], and End-of-Life (EoL) modelling scenarios [30]; Takano [31]. Clark [32] studied embodied equivalent carbon values for commercial buildings based on different methodologies and demonstrated results ranging from 300 to 1650 $kgCO_2eq/m^2$. De Wolf et al. [22] comprehensively investigated discrepancies in the final results of LCA studies due to the quality of available data.

The other factor contributing to further discrepancies that has been studied comprehensively in the literature is the adoption of LCI techniques. The LCI techniques include process, input–output, and hybrid methods. There are fundamental differences between the way data is treated and analysed in these techniques with the process analysis formed around disintegrating the relevant life cycle stages into characteristic processes. The data associated with each stage is collected directly from relevant manufacturers or is provided by specialist data inventories including ecoinvent and GaBi. The Input–output technique is formulated around financial transaction matrices between engaged sectors. The embodied energy values are calculated using energy intensity values that have been assigned to each sector. Hybrid analysis in designed to benefit from advantages of the two techniques and at the same eliminate their shortcomings [15]. The most impactful shortcomings of the two techniques include the 'truncation error' which is believed to significantly underrepresent requirements for the process analysis [33–37] and also the 'aggregation error' for input–output analysis for allocating similar energy intensity measures to all products within a sector [38].

There are various studies that have highlighted the discrepancies in embodied energy results associated with adoption of different LCI techniques. Crawford [39], Stephan et al. [15] and Stephan and Stephan [40] have demonstrated in their studies of whole buildings that a hybrid LCI analysis can lead to embodied energy values of up to four times greater than those achieved using a process analysis. In a similar study Wiedmann et al. [41] explored the environmental impact of wind turbines and demonstrated twice as high environmental impacts for hybrid analysis compared with a process analysis. Bontinck et al. [42]. A hybrid LCI was used to explore structural insulated panel systems. The findings of the hybrid analysis were found to be 159 percent higher than a process analysis and 46 percent lower than an input-output analysis. Guan et al. [43] conducted a process study on a hybrid LCA of a building in China and found a 100 percent gap.

Drastic disparities of this nature highlight the need for a harmonised and standardised LCA to be adopted by building regulations allowing for an effective decision-making tool to assist in the early stages of building fabric design, or strategising future policies and product development for various stakeholders.

Gelowitz and McArthur [44] reviewed the published EPDs for building products and identified adoption of different LCI methodologies, high level of incomparability between EPDs using the same PCR, and poor verification practices as the main barriers in adopting the results in further analysis. Although conceding that the number of valid comparisons were substantially greater for EPD generated in compliance with EN 15804, Resalati et al. [1] argue that the EN 15804 harmonisation standard has not been totally effective.

Although several studies have investigated the LCA concept in detail and provided insight into how best these tools could be further optimised for decision making processes, their use is currently primarily limited to academic studies [30], and is not incorporated in the industrial ecosystems in enough depth [7]. This has been attributed to a series of factors in the literature including the lack of appropriate interoperability between LCA methodologies and high demand tools in the construction sector

(Anand and Amor [30], Means and Guggemos [45]), the expertise required to carry out LCA studies reliably [45], and LCA priority for various industries at present in parallel to the confidentiality issues the manufacturers see as a barrier in publishing their LCA results [46]. As noted by Resalati et al. [1], such challenges may cause delays in the adoption of such technologies, implying that environmental policies and many of the assumptions on which current policies are founded may not accurately reflect energy and its consequent carbon investments. Several researchers such as Chastas et al. [47], Cellura et al. [48] and Moran et al. [49] have questioned whether our current energy efficiency measures with a focus on 'operational energy only', instead of a 'total energy' efficiency, are acceptable in the context of longer term strategic policy making.

This chapter, takes on an aggregated operational and embodied energy approach, aiming to demonstrate the impact of the uncertainties of embodied energy data when achieving low and zero energy buildings. The analyses aim to apply the aggregated approach on individual building elements and materials.

This chapter seeks to highlight the significance of considering the uncertainties of embodied energy data when LCA is used as decision making tool to inform the engaged stakeholder and other relevant end users. This will be carried out with a particular view of individual building components and materials, based on a total energy/carbon analysis.

This is illustrated by examining the sensitivity of optimal building insulation level to the deviations of embodied energy data. The assessments are shown in the context of residential buildings in the United Kingdom, although the methodology is not restricted to that and may be applied to a broader operational setting.

3. Aggregated embodied and operational carbon

While the connection between U-values and operational energy/carbon is generally linear (**Figure 2**), embodied carbon tends to rise at a faster rate as buildings are insulated to better efficiencies. Only thermal conductivity is positively associated with operational carbon (i.e. the line is not dependant on the insulation type).

Embodied carbon however, is directly dependent on the type of insulation and increases in the level of insulation progressively increase the embodied carbon values relative to thermal conductivity of the material and its associated embodied carbon burden.

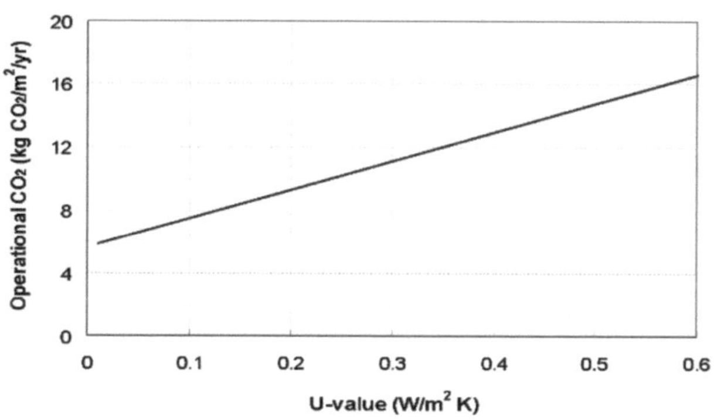

Figure 2.
Linear relationship between operational carbon and U-value.

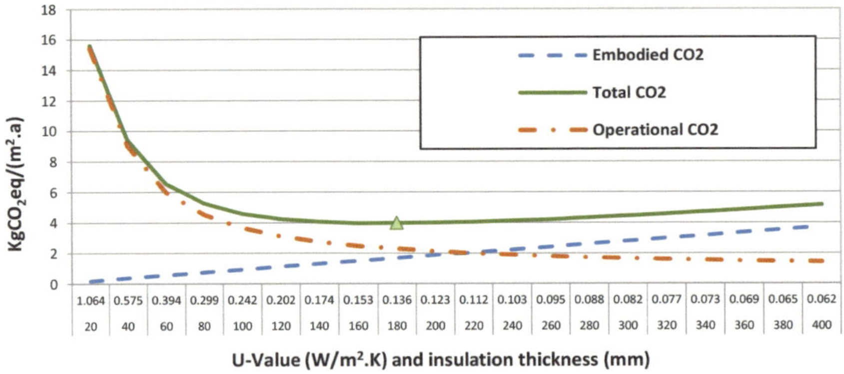

Figure 3.
Total carbon curve.

Figure 3 illustrates the total carbon curve (for PUR insulation as an example) for a typical dwelling. The key feature is that the aggregated total of the linear and non-linear relationship is inevitably non-linear. The graph demonstrates a progressively diminishing return for incremental improvement in U-value measures. Reducing the total carbon value therefore becomes more challenging to achieve using the building fabric insulation levels.

The graph indicates an optimum thickness for the insulation level beyond which the additional embodied carbon investment cannot be recovered through operational carbon savings (the marked point on **Figure 3**). The optimal point can change as the base assumptions are adjusted in the analysis e.g., insulation type, climate, occupancy levels, etc. The key feature however is that the three lines form a curve that repeats in all comparable scenarios. Such curves will eventually flatten for a longer service life or the use of an insulation material with lower associated embodied carbon. Such analyses demonstrate where optimum net benefit is achieved. Beyond these optima embodied carbon burdens exceed operational savings, whilst in advance of these points embodied carbon investment usefully reduced operational requirements.

This form of analysis is key when it comes to designing low/zero energy buildings where the existing standards tend to move towards even lower U-value requirements. On a material level, the analysis demonstrates that many conventional insulation materials cannot achieve very low U-values without incurring carbon disbenefits, whilst other conventional or novel materials with lower embodied carbon relative to their thermal conductivities, can justifiably achieve ambitious U-values.

The flat nature of the total carbon curve naturally creates comparative points on the graph, where lower levels of insulation show parity to the more extreme measures. The identified areas on **Figure 4** are referring to the insulation levels that are within 5% variation of the sweet spot i.e., the total carbon level associated with the 300mm insulation is identical to that of 100mm, in this specific case, but within 5% similarity to the total carbon level on the sweet spot. The operational only approach suggests 50% savings for the same range. This is crucial to be incorporated in all future building design strategies if the lower emission targets are to be met where a decarbonised grid coupled with electricity dominated operational energy demand is in the horizon.

Such findings will have significant financial implications as well for building design where higher levels of insulation would be more difficult to justify in the future zero energy building codes and standards. This similarly applies to setting the energy efficiency targets for retrofitting the existing building stock around the world.

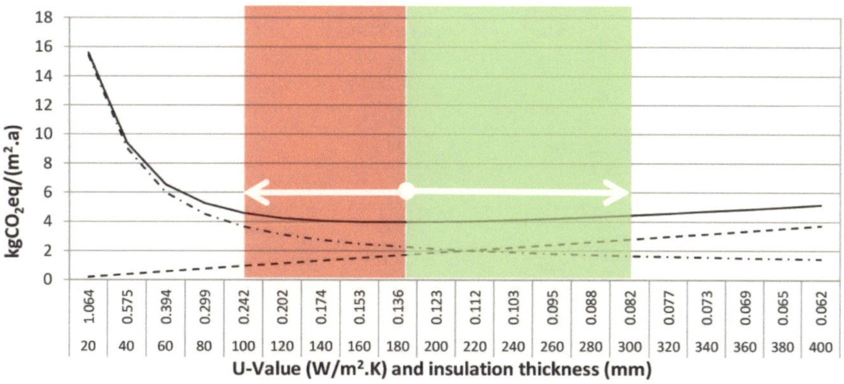

Figure 4.
Areas on the total carbon curve where lower levels of insulation show parity to the higher levels.

It is important to realise that the optimal points on the total carbon curve may move towards lower or higher insulation levels depending on the occupancy patterns, climatic conditions, building function and service life, source of energy, HVAC type and settings, and the type of insulation used, but the approach will be valid and its key feature still applicable, as identified above.

In order to demonstrate the extent of variability in results, the following section demonstrate the application of a series of insulation materials on a case study building in the UK. The assessments were conducted on a three-bedroom semi-detached house built in compliance with the most recent Building Regulations in the United Kingdom, as outlined in L1A Conservation of fuel and power. The building has a total floor area of 80 m^2.

4. Thermal and environmental performance of insulation materials

The studied insulation materials and their associated reported embodied carbon values are presented in **Table 2**. The values are extracted from available EPDs for each insulation material.

A box and whisker plot was used to graphically illustrate the locality and spread of GWP/unit weight data extracted from EPD documents as presented in **Table 2**. The interquartile range (IQR) between the 25th and 75th percentiles, median and standard deviation values are represented with the box, the line, and the whiskers respectively.

A clear distinction can be observed between studied insulation materials with MW and GW presenting comparable median values. EPS, XPS, PU, and PF as insulation materials with a hydrocarbon base also form a distinctive group with very similar median points. The PU and MW however are demonstrating greater probability distribution due to the discrepancies in the data compared with PF and GW.

For a more meaningful comparison between the insulation materials, their relative thermal performance needs to be reflected in the analysis. The thermal resistance (R-value) target for this analysis was taken as 6.6 m^2.K/W complying with the UK Building Regulation requirements as explained above. **Figure 6** represents the values in **Table 2** converted into the GWP values associated with the target thermal resistance for each insulation material.

The distinct insulation groups, with similar median points, that were formed in **Figure 5** are not evident in **Figure 6** and the GWP values in relation to the thermal

Material	f.u.	Density (kg/m^3)	λ (W/m k)	GWP ($kgCO_2eq/kg$)
Glass wool (GW)	m^3	15	0.0425	1.07
	m^3	19	0.0395	1.07
	m^3	24	0.035	1.16
	m^3	31	0.033	1.07
	m^3	10.5	0.044	1.16
	m^3	19.5	0.035	0.97
	m^3	11.5	0.04	0.99
	m^3	20	0.035	1.43
Mineral wool (MW)	m^2	38.5	0.03676	0.85
	m^3	33	0.039	1.63
	m^3	85	0.04	1.13
	m^3	50	0.035	1.53
	m^2	29	0.037	1.21
	m^3	41	0.04	0.84
	m^3	94	0.04	0.88
	m^3	158	0.04	0.89
	m^3	20.5	0.036	1.24
	m^3	23.5	0.0335	1.81
	m^3	22.3	0.0335	1.86
	m^3	14.8	0.04	1.72
	m^3	15	0.04	1.92
Expanded polystyrene (EPS)	m^3	15.5	0.035	2.99
	m^2	25	0.034	2.35
	m^3	15.5	0.035	2.99
	m^3	15.5	0.035	2.99
	m^3	22.5	0.035	3.51
	m^3	16.6	0.035	2.89
	m^3	22.9	0.035	2.71
Extruded polystyrene (XPS)	m^2	35	0.031	2.91
	m^2	33.7	0.035	2.79
	m^2	34.6	0.035	2.75
	m^2	33.7	0.035	2.79
Polyurethane (PU)	m^2	31	0.023	4.03
	m^2	31	0.023	3.47
	m^2	31	0.026	3.52
	m^2	31	0.026	3.20
	m^2	40	0.026	3.19
	m^2	42	0.023	7.76
	m^2	32	0.023	3.81

Material	f.u.	Density (kg/m^3)	λ (W/m k)	GWP (kgCO_2eq/kg)
Phenolic foam (PF)	m^2	35	0.021	2.83
	m^2	35	0.021	2.91
Foam glass (FG)	m^3	165	0.103	0.12
	m^3	130	0.082	0.12
	kg	117	0.041	1.30
Cellulose (CEL)	m^3	28	0.039	0.13
	m^3	28	0.039	0.10
	kg	31	0.039	0.20
	m^3	250	0.049	0.86
	m^3	160	0.04	0.64
	m^3	260	0.05	0.70
	m^3	140	0.038	0.43
	m^3	240	0.047	0.44
	m^3	210	0.044	0.39
Vacuum insulation panel (VIP)	kg	200	0.007	9.4
	kg	200	0.007	11.1
	kg	200	0.007	6.4

Table 2.
Environmental properties of insulation materials with reference to their thermal conductivities.

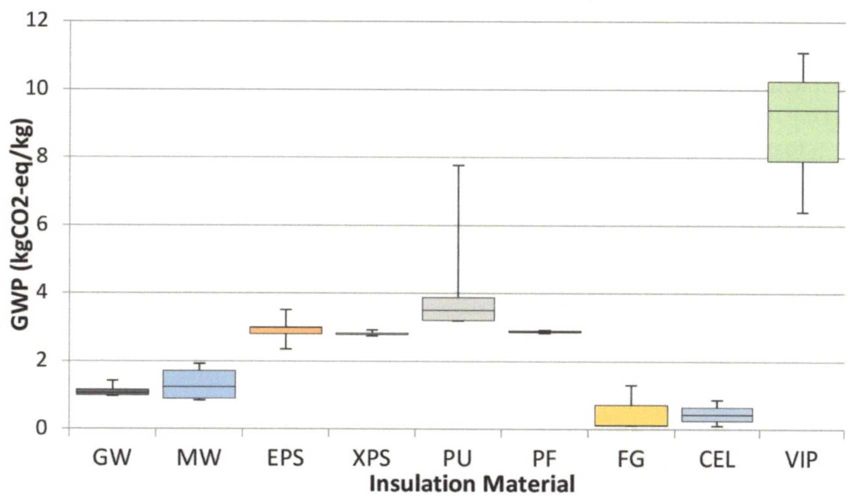

Figure 5.
GWP per unit weight of different insulation materials.

resistance demonstrate smaller variation between the insulation groups. The distribution of GWP values for MW becomes significantly broader when the targeted R-value is considered, whereas these values for GW stays relatively stable.

The median point for XPS insulation material demonstrates 100% higher GWP values compared with EPS. The GWP values for Cellulose-based insulations, cover

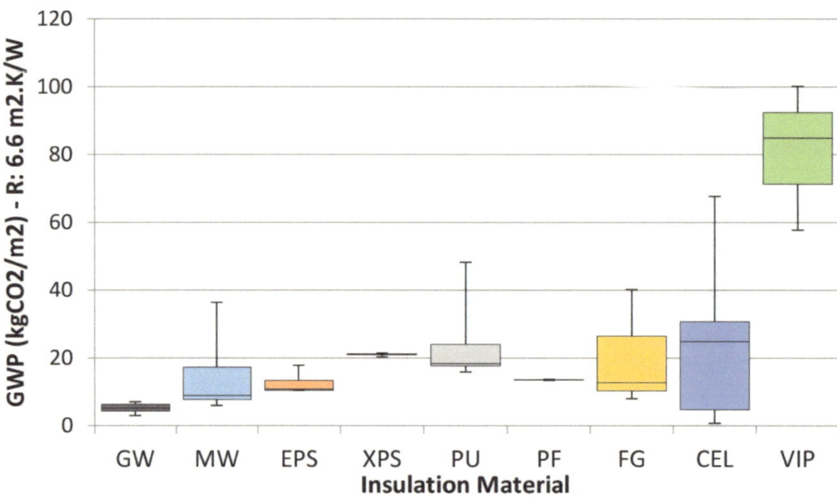

Figure 6.
GWP per unit area of insulation materials for the target thermal resistance.

a broad range from 0.7–67 $kgCO_2$_eq/m^2. This is due to the variety of base products used for making cellulose-based insulations. The base products can include refined virgin wood chips and blown recycled cellulosic products, such as wastepaper with requiring their own dedicated processing procedures.

The GWP values for VIPs are significantly higher than the other insulation materials even with factoring their considerably better thermal performance (up to 10 times better) in the analysis. This must be noted however that over 90% of the GWP values associated with VIPs are linked to their core material and specifically in the studied EPD associated with the use pyrogenic silica [50, 51]. The studied EPDs for VIPs are based on a cradle to gate approach and therefore not taking into account the recyclability potential for VIPs. Considering different end-of-life scenarios could change the impact of VIPs and other insulation materials such as cellulose based materials significantly.

5. Optimum U-value levels on the basis of a total carbon analysis

The following section utilises the range of embodied carbon values presented above in order to identify the associated optimum U-value measures and present the uncertainties such discrepancies can cause in early stage building design decision making. **Figure 7** demonstrates these points and clearly presents the broad range of identified optimum points that can be achieved using the same type of insulation material depending on the source of embodied carbon values used.

The range of embodied carbon associated with Cellulose insulation as an example, leads the optimum U-value points to cover values from 0.15 W/m^2.K to 0.35 W/m^2.K. This applies to all other insulation materials as well with MW and PU covering 0.16–0.25 W/m^2.K and 0.21–0.29 W/m^2.K respectively.

Comparing VIP values with other insulation materials demonstrate that U-values lower than 0.21 W/m^2.K could not be reached without leading to an increase in the total carbon values. The VIP values however show comparable results with PU insulation, although due to its higher GWP values, the total carbon value is higher for identical optimum U-values. This is also investigated by Resalati et al. [1] where it was

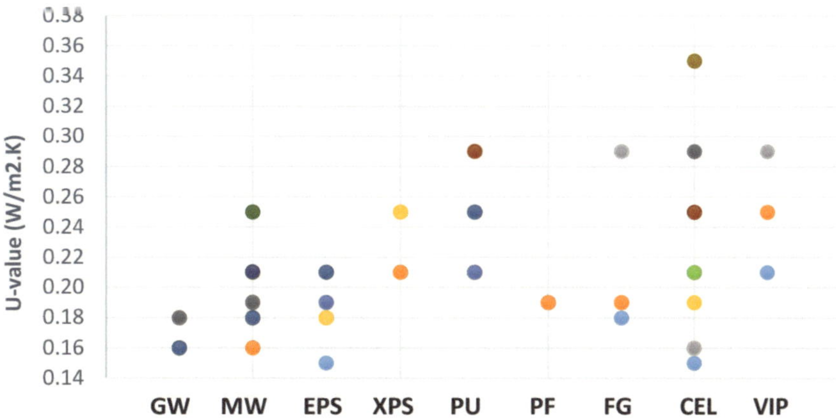

Figure 7.
Optimum U-value points associated with the GWP data points generated on the total carbon curve.

observed for VIPs that their interquartile range was almost double of those for PU. The values demonstrate the CEL, EPS, GW, and MW insulation types allow for lower U-values to be reached, in the context of assumptions applied to this study. **Figure 7** further highlights the sensitivity of identifying the optimum insulation levels for low and zero energy buildings to the assumption applied to the LCA models.

The optimum insulation levels based on an aggregated operational and embodied carbon approach allows for identifying the effectiveness of building fabric design in meeting the carbon savings targets. This has been presented that an operational carbon only approach, as is required by the current building energy codes and regulations, does not necessarily lead to a lower overall energy load when compared with an aggregated approach. This has also been concluded by Mohazabieh et al. [52], Gul and Patidar [53] and Stephan et al. [15].

The analyses here further highlights these implications for subsequent future regulatory requirements, and hence provide the building product manufacturers with appropriate tools for analysing their products' place in any future market where a total carbon approach is applied to building design in principle. The key concept here is that the discussion of factoring embodied energy/carbon into building design decisions is well past the point of questioning its significance and more addressing the challenges of how best this could be incorporated into our existing regulations. This is also concluded in a study presented by Lutzkendorf [54]. Further delays in factoring in the environmental performance of various materials, products, and services when calculating/regulating the required U-values in building design can in principle lead to the design choices that increase the carbon footprint rather than reducing it.

The findings also provide meaningful insight for developing novel insulation technologies. Any new technology will need to have very low levels of embodied energy relative to its R-value if lower insulation levels are to be achieved with the embodied energy values factored in. This can either be achieved using low impact materials or with appropriate plans for end-of-life recyclability. VIPs for instance offer huge potential to be used in future low and zero energy buildings given their very thin nature relative to the thermal conductivity measures, and can potentially outperform the conventional insulation materials based on an aggregated carbon approach. Appropriate end of life treatments however, need to be considered for VIPs to be competitive in the market environmentally.

6. Conclusions

Although existing assessment techniques have specific shortcomings, the number of research and initiatives currently undertaken in many countries highlight that the Life Cycle Assessment of buildings will be a feature of future assessments of building environmental impacts. Regulations incentivizing additional stakeholders to use these methodologies, as recommended by Eurima, should be the driving force behind increased acceptance of assessments of this type [7]. This adoption, on the other hand, requires studies that can deliver practical roadmaps, supporting the engaged stakeholders in establishing effective and long term business strategies. A more in depth understanding of the restrictions of LCA studies is a necessary requirement for developing reliable methodologies that can deliver high accuracy and reliability in a practical way.

When aggregated operational and embodied carbon are taken into account, total carbon curves have been formed identifying sweet spots where the embodied carbon investment cannot be recovered through operational savings. Data uncertainties, occupancy patterns, climatic conditions, building function and service life, source of energy, HVAC type and settings, and the type of insulation used, all contribute to the theoretical minimum. As a result, identical optimal specifications cannot be provided for various scenarios, rather, sufficient analytical and predictive understanding is required.

Considerable total energy savings can be achieved by practices and standards based on such principles. Such studies can help determine optimum insulation levels that can be incorporated into design of a building or that may be needed by standards in the future, as well as the limits to how much present energy-saving methodologies can be increased using certain technologies. Although whole life cycle thinking is now acknowledged in several codes and standards around the world, the lack of availability of reliable and accurate data, and the differences in adopting the existing methodologies for generating EPDs and other LCA results can lead to generating misleading messages to the manufacturers and policymakers. The study provides evidence in favour of better harmonisation and standardisation of LCA and LCI databases and procedures.

Due to high variation in the LCA results as a result of current discrepancies in modelling assumptions, applied methodologies, and data, the total carbon approach can be utilised as guideline for the time being, while the onus remains on LCA specialists and practitioners, as well as other key stakeholders, to harmonise the science across all industries, including software.

Author details

Shahaboddin Resalati
Oxford Brookes University, Oxford, UK

*Address all correspondence to: sresalati@brookes.ac.uk

References

[1] Resalati S, Kendrick CC. Hill C. Embodied energy data implications for optimal specification of building envelopes. Building Research & Information. 2020;**48**(4):429-445

[2] Anastaselos D, Giama E, Papadopoulos AM. An assessment tool for the energy, economic and environmental evaluation of thermal insulation solutions. Energy and Buildings. 2009;**41**(11):1165-1171

[3] Häfliger IF, John V, Passer A, Lasvaux S, Hoxha E, Saade MRM, et al. Buildings environmental impacts' sensitivity related to LCA modelling choices of construction materials. Journal of Cleaner Production. 2017;**156**:805-816

[4] Koezjakov A, Urge-Vorsatz D, Crijns-Graus W, van den Broek M. The relationship between operational energy demand and embodied energy in Dutch residential buildings. Energy and Buildings. 2018;**165**:233-245

[5] Ramesh T, Prakash R, Shukla KK. Life cycle energy analysis of buildings: An overview. Energy and Buildings. 2010;**42**(10):1592-1600

[6] Szalay AZZ. What is missing from the concept of the new European Building Directive? Building and Environment. 2007;**42**(4):1761-1769

[7] Eurima. Life Cycle Assessment of Buildings: A Future-proofed Solution in the Digitalised World of Tomorrow. [ebook]. 2017. Available from: https://www.eurima.org/uploads/ModuleXtender/Publications/170/Eurima_LCA_WhitePaper_Final_20170915.pdf

[8] Tettey UYA, Dodoo A, Gustavsson L. Effects of different insulation materials on primary energy and CO_2 emission of a multi-storey residential building. Energy and Buildings. 2014;**82**:369-377

[9] Cole & Fedoruk. Shifting from net-zero to net-positive energy buildings. Building Research & Information. 2015;**43**(1):111-120. DOI: 10.1080/09613218.2014.950452

[10] RICS. Methodology to calculate embodied carbon of materials, RICS information paper, IP32/2012. Coventry, UK: Royal Institution of Chartered Surveyors (RICS). 2012

[11] Kristjansdottir TF, Heeren N, Andresen I, Brattebø H. Comparative emission analysis of low-energy and zero-emission buildings. Building Research & Information. 2018;**46**(4):367-382. DOI: 10.1080/09613218.2017.1305690

[12] Sartori I, Hestnes AG. Energy use in the life cycle of conventional and low-energy buildings: A review article. Energy and Buildings. 2007;**39**(3):249-257

[13] Dixit MK. Embodied energy and cost of building materials: Correlation analysis. Building Research & Information. 2017;**45**(5):508-523

[14] Chau CK, Leung TM, Ng WY. A review on life cycle assessment, life cycle energy assessment and life cycle carbon emissions assessment on buildings. Applied Energy. 2015;**143**:395-413

[15] Stephan A, Crawford RH, De Myttenaere K. A comprehensive assessment of the life cycle energy demand of passive houses. Applied Energy. 2013;**112**:23-34

[16] Dascalaki EG, Argiropoulou PA, Balaras CA, Droutsa KG, Kontoyiannidis S. Benchmarks for embodied and operational energy assessment of Hellenic Single-Family Houses. Energies. 2020;**13**(17):4384

[17] Mourao J, Gomes R, Matias L, Niza S. Combining embodied and operational energy in buildings refurbishment assessment. Energy and Buildings. 2019;**197**:34-46

[18] Gustavsson L, Joelsson A. Life cycle primary energy analysis of residential buildings. Energy and Buildings. 2010;**42**(2):210-220

[19] Azari R, Abbasabadi N. Embodied energy of buildings: A review of data, methods, challenges, and research trends. Energy and Buildings. 2018;**168**:225-235

[20] Dascalaki EG, Argiropoulou P, Balaras CA, Droutsa KG, Kontoyiannidis S. Analysis of the embodied energy of construction materials in the life cycle assessment of Hellenic residential buildings. Energy and Buildings. 2021;**232**:110651

[21] Takano A, Hughes M, Winter S. A multidisciplinary approach to sustainable building material selection: A case study in a Finnish context. Building and Environment. 2014;**82**:526-535

[22] De Wolf C, Pomponi F, Moncaster A. Measuring embodied carbon dioxide equivalent of buildings: A review and critique of current industry practice. Energy and Buildings. 2017;**140**:68-80

[23] Lotteau M, Loubet P, Pousse M, Dufrasnes E, Sonnemann G. Critical review of life cycle assessment (LCA) for the built environment at the neighborhood scale. Building and Environment. 2015;**93**:165-178

[24] Soust-Verdaguer B, Llatas C, García-Martínez A. Simplification in life cycle assessment of single-family houses: A review of recent developments. Building and Environment. 2016;**103**:215-227

[25] Moncaster AM, Song J-Y. A comparative review of existing data and methodologies for calculating embodied energy and carbon of buildings. International Journal of Sustainable Building Technology and Urban Development. 2012;**3**(1):26-36. DOI: 10.1080/2093761X.2012.673915

[26] Hill C, Zimmer K. The Environmental Impacts of Wood Compared to Other Building Materials. Norway: Norwegian Institute of Bioeconomy Research (NIBIO); 2018

[27] Cabeza LF, Rincón L, Vilariño V, Pérez G, Castell A. Life cycle assessment (LCA) and life cycle energy analysis (LCEA) of buildings and the building sector: A review. Renewable and Sustainable Energy Reviews. 2014;**29**:394-416

[28] Rauf A, Crawford RH. Building service life and its effect on the life cycle embodied energy of buildings. Energy. 2015;**79**:140-148

[29] Silvestre JD, De Brito J, Pinheiro MD. Environmental impacts and benefits of the end-of-life of building materials–calculation rules, results and contribution to a "cradle to cradle" life cycle. Journal of Cleaner Production. 2014;**66**:37-45

[30] Anand CK, Amor B. Recent developments, future challenges and new research directions in LCA of buildings: A critical review. Renewable and Sustainable Energy Reviews. 2017;**67**:408-416

[31] Frischknecht R. LCI modelling approaches applied on recycling of

materials in view of environmental sustainability, risk perception and eco-efficiency. The International Journal of Life Cycle Assessment. 2010;**15**(7):666-671

[32] Clark DH. What Colour is Your Building?: Measuring and Reducing the Energy and Carbon Footprint of Buildings. London: RIBA Publishing; 2013

[33] Crawford RH. Validation of a hybrid life-cycle inventory analysis method. Journal of Environmental Management. 2008;**88**(3):496-506

[34] Crawford RH, Bontinck PA, Stephan A, Wiedmann T. Towards an automated approach for compiling hybrid life cycle inventories. Procedia Engineering. 2017;**180**:157-166

[35] Crawford RH, Bontinck PA, Stephan A, Wiedmann T, Yu M. Hybrid life cycle inventory methods: A review. Journal of Cleaner Production. 2018;**172**:1273-1288

[36] Lenzen M. Errors in conventional and Input-Output—based Life—Cycle inventories. Journal of Industrial Ecology. 2000;**4**(4):127-148

[37] Majeau-Bettez G, Strømman AH, Hertwich EG. Evaluation of process-and input–output-based life cycle inventory data with regard to truncation and aggregation issues. Environmental Science & Technology. 2011;**45**(23): 10170-10177

[38] Säynäjoki A, Heinonen J, Junnila S, Horvath A. Can life-cycle assessment produce reliable policy guidelines in the building sector? Environmental Research Letters. 2017;**12**(1):013001

[39] Crawford R. Life Cycle Assessment in the Built Environment. London, UK: Routledge; 2011

[40] Stephan A, Stephan L. Reducing the total life cycle energy demand of recent residential buildings in Lebanon. Energy. 2014;**74**:618-637

[41] Wiedmann TO, Suh S, Feng K, Lenzen M, Acquaye A, Scott K, et al. Application of hybrid life cycle approaches to emerging energy technologies—The case of wind power in the UK. Environmental Science & Technology. 2011;**45**(13):5900-5907

[42] Bontinck PA, Crawford RH, Stephan A. Improving the uptake of hybrid life cycle assessment in the construction industry. Procedia Engineering. 2017;**196**:822-829

[43] Guan J, Zhang Z, Chu C. Quantification of building embodied energy in China using an input–output-based hybrid LCA model. Energy and Buildings. 2016;**110**:443-452

[44] Gelowitz MDC, McArthur JJ. Comparison of type III environmental product declarations for construction products: Material sourcing and harmonization evaluation. Journal of Cleaner Production. 2017;**157**:125-133

[45] Means P, Guggemos A. Framework for life cycle assessment (LCA) based environmental decision making during the conceptual design phase for commercial buildings. Procedia Engineering. 2015;**118**:802-812

[46] Han G, Srebric J. Comparison of survey and numerical sensitivity analysis results to assess the role of life cycle analyses from building designers' perspectives. Energy and Buildings. 2015;**108**:463-469

[47] Chastas P, Theodosiou T, Bikas D, Kontoleon K. Embodied energy and nearly zero energy buildings: A review in residential buildings. Procedia Environmental Sciences. 2017;**38**:554-561

[48] Cellura M, Guarino F, Longo S, Mistretta M. Energy life-cycle approach in Net zero energy buildings balance: Operation and embodied energy of an Italian case study. Energy and Buildings. 2014;**72**:371-381

[49] Moran P, Goggins J, Hajdukiewicz M. Super-insulate or use renewable technology? Life cycle cost, energy and global warming potential analysis of nearly zero energy buildings (NZEB) in a temperate oceanic climate. Energy and Buildings. 2017;**139**:590-607

[50] Karami P, Al-Ayish N, Gudmundsson K. A comparative study of the environmental impact of Swedish residential buildings with vacuum insulation panels. Energy and Buildings. 2015;**109**:183-194

[51] Schonhardt U, Binz A, Wohler M, Dott R. Ökobilanz eines Vakuum-Isolations-Paneels (VIP). Basel, German: University of Applied Sciences, Institute of Energy; 2003

[52] Mohazabieh SZ, Ghajarkhosravi M, Fung AS. Energy consumption and environmental impact assessment of the energy efficient houses in Toronto, Canada. Procedia Engineering. 2015;**118**:1024-1029

[53] Gul MS, Patidar S. Understanding the energy consumption and occupancy of a multi-purpose academic building. Energy and Buildings. 2015;**87**:155-165

[54] Lützkendorf T. Assessing the environmental performance of buildings: Trends, lessons and tensions. Building Research and Information. 2018;**46**:594-614. DOI: 10.1080/09613218.2017.1356126

Chapter 3

Coalash as Sustainable Material for Low Energy Building

Avijit Ghosh

Abstract

Sand, which is a naturally occurring soft mineral ranks second after water, as far as consumption is concerned globally. Due to rapid infrastructural development worldwide, particularly in Asian region, the rate of natural formation of sand has been found to be outpaced by rate of consumption, causing greater ecological imbalances. Coalash, an industrial waste from thermal power plants are polluting in nature, and legacy ash in huge proportion without proper utilization is posing a serious threat to the environment. It was ideated to replace sand by coalash in concrete and mortar mix, and to evaluate the physical and thermal properties for its suitability in low energy building construction. Without compromising strength criteria, thermal transmittance value is found to be reduced up to considerable extent, which resulted lesser cooling requirement with added economic benefit. This medium technology application could be one of the economic pathway towards Near Zero Building Construction.

Keywords: sand, coal ash, energy, envelop, building

1. Introduction

Coal, being the largest source of power worldwide, contributes maximum in respect of global CO_2 Emission. More than 33% of global electricity was produced by 2100 GW cumulative capacity coal based thermal power plants in 2020. China, United States and India are ranked as top three nations for coal combustion linked electricity production [1]. This being fossil fuel, and linked to CO_2 emission, worldwide movement has been started to tone down its usage. The ash produced as a fallout of coal burning is classified into fly ash and bottom ash. It is considered that for every 4 tonnes of coal burnt, approximately one ton of coal ash is produced [2]. The ash content actually depends on type of coal burnt, which is classified as anthracite, bituminous, sub-bituminous and lignite. But, the legacy ash already stored in various ash ponds of thermal power plants is huge and continuously affecting the environment by contaminating surrounding air and soil. In spite of various efforts undertaken by authorities, 100% ash utilization could not be made possible. On the other hand, sand, the natural soft mineral is continuously depleted due to mindless sand mining mainly for infrastructure related developmental requirements. In fact, rate of formation of sand is getting outpaced by rate of extraction, thus creating a serious environmental imbalances. Considering the rapid growth in urbanization and particularly huge demand in building sector, the energy consumption by building sector alone is enormous. Overall, buildings accounted for 36 per cent of global energy demand and 37 per cent of energy-related CO_2 emissions in 2020. Since the signing of the Paris Agreement in 2015, greenhouse gas emissions from the buildings and construction sector had peaked in 2019 and subsequently fallen to 2007 levels in 2020 mostly due to the COVID-19 pandemic [3]. Despite the expected rebound in emissions in 2021 being moderated by continued power sector de-carbonization, buildings remain off track to achieve carbon neutrality by 2050. To meet this target, all new buildings and 20% of the existing building stock would need to be zero-carbon-ready as soon as 2030 [4]. Buildings consume energy at different levels of the life cycle. The fastest-increasing end uses of energy in buildings are for space cooling, appliances and electric plug-loads, which contribute buildings sector electricity demand growth. Researchers observed that operational energy requirement by buildings occupy lion's share (around 80%) and rest is shared by material embodied energy including transportation, construction etc. related energy consumptions. It was also observed that a normal residential purpose use building and an office purpose use building consume on an average 275 $kWh/m^2/year$, and 400 $kWh/m^2/year$ respectively [5]. Though these figures depend on many factors like climatic condition of the area, material choice, orientation / layout etc. the overall life-cycle energy figure can be optimized by appropriate planning and design. Among all the building components, envelop influences in deciding the energy ingress or egress to and from the building core. Such flow of energy ultimately determines the operational energy requirement for the particular building, depending on its functionality. Therefore the design of envelop with appropriate material property can contribute significantly from energy efficiency point of view. This research topic presents an approach in passive design of building envelop by selecting waste based abundantly available material, which might evolve into less heat conducting concrete and mortar. Thus, coal ash substitution in building construction industry address the issues of effective thermal power plant waste utilization, arresting rapid depletion of sand and improvement in building energy efficiency together.

2. Building envelop and constituent materials

2.1 Conventional materials

The building envelop influences heat conduction through roof, wall, fenestrations and determine the quantum of sensible cooling/heating load requirement to balance comfort condition. As per the report, published by International Energy Agency in 2013, the demand for space air-conditioning is estimated to rise three fold between 2010 and 2050 on account of more numbers of hot days [6]. To restrict more heat entry inside the building, insulating materials are put to use. The conventional materials, which are used for this purpose are mostly by-products of fossil fuel oil industries, and the cost and embodied energy content of those are very high, besides being hazardous at the end of life disposal scenario. Choice of materials for energy efficient envelop construction should cater to the issues of durability, environmental sustainability, local sourcing of materials to reduce transportation related emission etc. Concrete is an integral composite material in modern building construction with considerable carbon footprint, due to one of its constituent with high embodied energy content, which is cement. It has been observed that Ordinary Portland Cement (OPC) contains the highest Global Warming Potential (GWP), Photochemical Ozone

Creation Potential (POCP) and Abiotic Depletion Potential (ADP) [7]. Paints with high reflection parameter on roof and other types of insulating materials are included in energy efficient building design. The manufacturing energy and GWP of normally used such materials are on the higher side [8]. Bergey had presented in a Symposia about the comparative study of various commercially available insulating materials, among which XPS was observed to be with highest embodied global warming potential (GWP) [9]. High albedo coating with cool roof feature contain embodied energy to the tune of 23 MJ/m^2 of roof surface [10].

The walls in a building are conventionally made up of bricks, joined with mortar and covered by plaster on both sides, topped with paint and other finishing. The materials used for mortar and plaster are cement and sand of different proportions and grade (MM3/MM5 etc.) [11].

Since, major carbon intensive component in concrete is cement, sustainable concrete mixes had been adopted for this work with the inclusion of Portland Pozzolana Cement (PPC) (30% fly ash blended), stone aggregate, sand, water and fly ash /bottom ash / marble dust / lime dust.

2.2 Coal ash as constituent material in envelop construction

Coal ash is basically a combination of lighter fly ash (75–80%) and coarser bottom ash particles, produced out of coal combustion in thermal power plants with zero embodied energy content. Depending upon its CaO percentage in the composition, it is classified as Class C (with some cementatious property) or Class F (with pozzolanic property). Globally, 100% utilization of coal ash from all the thermal power plants has not become possible till date. Cumulative accumulation of the un-utilized coal ash each year in ash dykes are creating groundwater contamination, air pollution etc. On the other hand, due to the rapid growth in global infrastructure sector, unprecedented rate and pace of sand mining from river bed is threatening the ecological balance enormously. Various researchers have explored the suitability of this industrial waste, which is coal ash in building construction, as a constituent material of concrete and mortar. To name a few, Higgins had compared one tone of concrete made of ordinary Portland cement as main constituent and the same quantity of concrete with Portland pozzolana cement with 30% flyash blend. It was observed that 17% less CO_2 emission to the atmosphere, 14% less primary energy requirement and 4% less mineral extraction resulted with such substitution [12]. The earlier works related to utilization of coal bottom ash in concrete have been studied. OPC, sand, bottom ash and stone aggregate as the concrete mix constituents were used. The results revealed that 10–30% replacement of sand by bottom ash did not adversely impact the desired strength gain in the concrete, barring some losses in workability and flexural strength parameters [13–19]. Another group of Researchers investigated about the suitability of fly ash and bottom ash as replacement material of cement and normal river sand utilized in concrete making. The compressive strength values at 28 days after casting were noted to be without change in comparison with conventional concrete mix ingredients. The workability parameter of the concrete mix was noted to be stiff, but at a longer maturity period, strength increased considerably. Toxicity parameters and durability aspects including leaching procedure, sulfate and acid attack and elevated temperature effects on concrete blended with coal ash as substitute to cement and sand were also studied, and the test results did not reflect any adverse impact, and as such considered to be used as clean construction material [20, 21]. Other researchers explored about the usage of fly ash as fine aggregate in masonry mortar and found that up to a considerable replacement ratio, the fly ash blended mortar can be used [22]. Soheil Oruji, Nicholas A. Brake and others tried to see if the finely ground bottom ash can act as an alternative material to cement in mortar preparation. The fineness effect on workability as well as, on setting time were studied. Improvement in micro-structure of cement mortar and increase in the strength parameter of such product was observed [23]. Kim had experimented with sieved and ground coal bottom ash in high strength cement mortar. The ground bottom ash was found to increase the workability and compressive strength values compared to the equivalent mortar made of cement and fly ash [24]. Shahidan et al. had studied the physical and chemical properties of coal bottom ash, as a replacement material for sand. The gradation of particles in bottom ash and sand showed some similarity, and overall, bottom ash is recommended favorably as a replacement material to sand [25]. Abbas et al. had also studied the effect on cement and sand by limestone dust and bottom ash partially respectively. For a number of mixes, sand substitution by bottom ash were done in various replacement ratios, and limestone dust replacement ratio with cement was maintained constant at 5% ratio. Water-cement ratio was same for all mixes. Increase in strength was found consistent up to 30% sand substitution and 5% cement substitution [26]. Ghosh et al. had experimented with coal bottom ash and fly ash separately as sand substitute in different concrete mix proportions. It was observed that with increasing percentage of replacement, thermal resistance parameter increased but the strength parameters decreased. Up to 40% replacement, the blended concrete exhibited desired strength with considerable percentage of decreased thermal conductivity value [27]. In another set of experiments, Ghosh et al. had further observed the effect of coal bottom ash and fly ash separately on masonry mortar of different proportions. The sand in the mortar was replaced by bottom ash and fly ash (separately) in steps of 10% up to 100%. The masonry mortar minimum strength criteria was observed to be fulfilled up to 100% replacement ratio, and specific mortar grade requirement was fulfilled up to 60% replacement with an astounding result of lower thermal conductivity [28].

3. Method

3.1 Materials and characterizations

In the research work, PPC, river sand, stone aggregate of 10 mm down size, potable quality water and coal ash were used. For material characterization, quantitative chemical analysis (for cement, sand, fly ash and bottom ash), X-Ray Diffractogram (XRD) (for cement, sand, fly ash and bottom ash), sieve analysis (for sand, bottom ash and stone aggregate), particle size analysis (for fly ash), density test (for cement, sand, fly ash, bottom ash and aggregate), surface area determination (cement, fly ash and bottom ash) and Finite Element Scanning Electron Microscopy (FESEM) (for sand, fly ash and bottom ash) were carried out as per standard testing protocol. Chemical analysis results are tabulated and XRD and FESEM images (of fly ash and bottom ash) are shown below (**Figures 1–4** and **Table 1**):

Grading curve obtained by sieve analysis for bottom ash sample and particle size curve of fly ash sample are shown below (**Figures 5** and **6**):

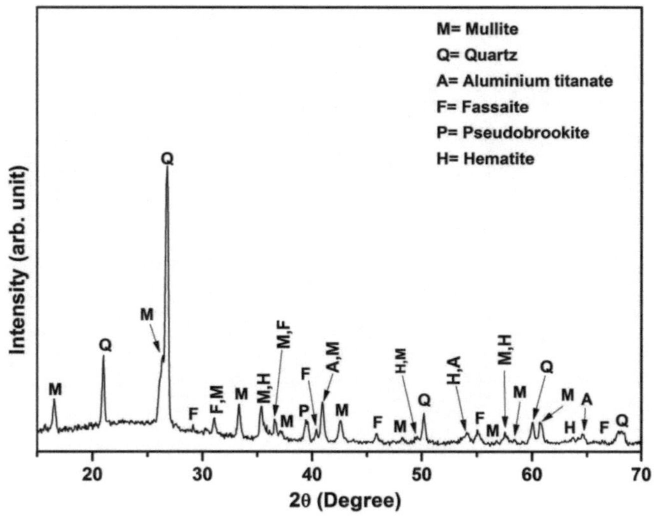

Figure 1.
XRD of fly ash sample.

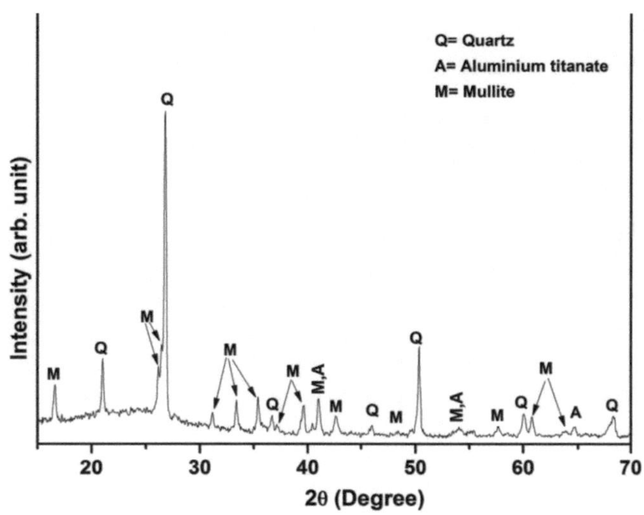

Figure 2.
XRD of bottom ash sample.

3.2 Experimental programme

The main objective of the experimental investigation is to ascertain the physical strength of the Concrete and Mortar mixes and finding out the thermal conductivity value of such mixes. The different mixes were designed with replacement of natural mineral by Coal ash and the changes thereof with respect to the physical and thermal properties.

Concrete mix design on the basis of basic ingredient material properties and fixing of proportions as per IS 456: 2000 [30], IS 10262: 2009 [31] and SP 23: 1982 [32] code provisions. Mortar mix selection as per relevant IS 2250: 1981 [11] code provisions.

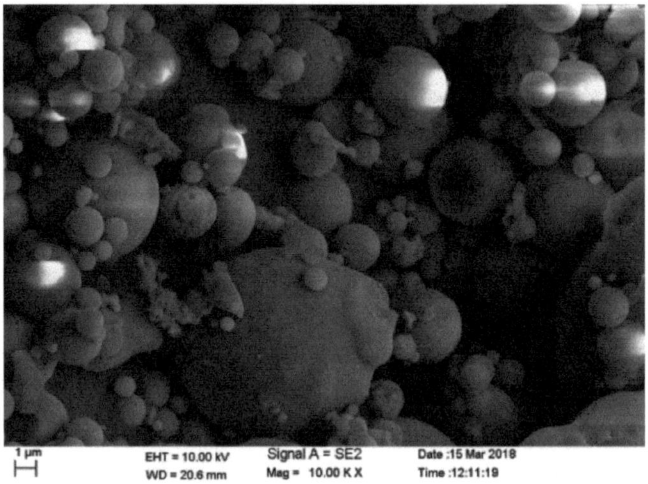

Figure 3.
FESEM of flyash at 10000X.

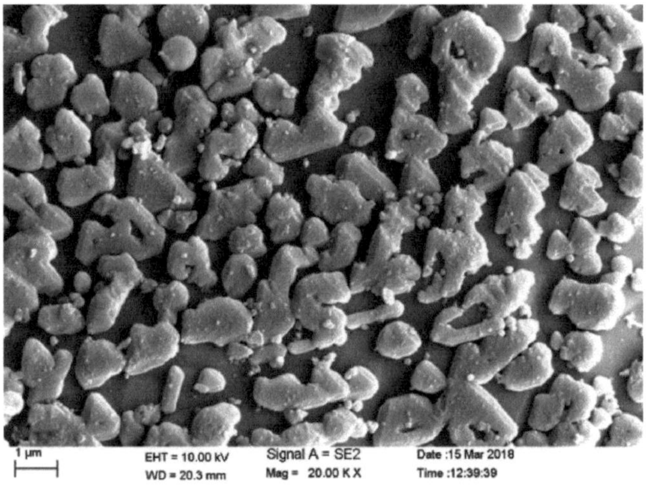

Figure 4.
FESEM of bottom ash at 20000X.

Universal testing machine for compressive strength determination and apparent porosity and bulk density test apparatus for concrete and mortar samples were utilized. For thermal conductivity determination of concrete and mortar samples, hot disk TPS 2500S instrument (working on Transient Plane Source method by following ISO 22007-2) was used and for overall heat transfer coefficient determination (U-value), guarded hot box method was adopted. Altogether around 200 samples were prepared and tested. Some of the concrete and mortar mixes are tabulated as below (**Tables 2–5**):

3.2.1 Guarded hot box test set-up to measure overall heat transfer co-efficient (U value)

A Guarded Hot Box test setup was designed and developed in School of Energy Studies, Jadavpur University, India. The setup was constructed following standard

Parameters tested	Requirement as per IS 3812	Test data of flyash used	Test data of bottom ash used
Silicon-di-oxide (SiO_2) + aluminum oxide (Al_2O_3) + iron oxide (Fe_2O_3) (%) by mass, Min.	70.00	51.38 + 33.12 + 6.87 = 91.37	60.71 + 25.86 + 6.81 = 93.38
Silicon di-oxide (SiO_2) (%) by mass, Min.	35.00	51.38	60.71
Magnesium oxide (MgO) (%)by mass, Max.	5.00	0.47	0.63
Total sulfur as sulfur tri-oxide (SiO_3) (%) by mass, Max.	2.75	0.09	0.15
Available alkalis as sodium oxide (Na_2O) % by mass, Max.	1.50	0.72	0.38
Loss on ignition, % by mass, Max.	12.00	1.80	0.92

Table 1.
Chemical composition of fly ash and bottom ash.

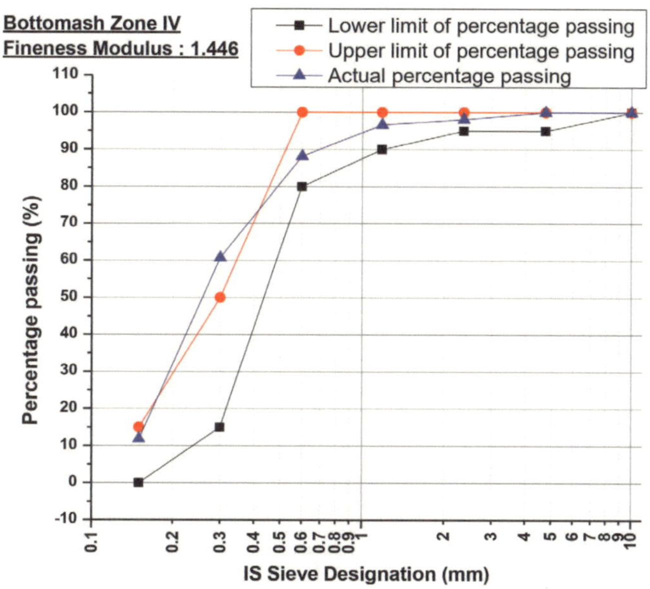

Figure 5.
Grading curve of bottom ash as per IS 383 by sieve analysis [29].

protocol. This setup was constructed using high insulating material, extruded polystyrene, whose thermal conductivity is 0.027 W/m K. This setup is capable to measure the U-value of any material whose thermal conductance is in the range of 0.1 W/m^2 K to 15 W/m^2 K. The testing of U-value of 125 mm thick burnt clay brick wall was done in both the cold side open and closed condition (**Figure** 7).

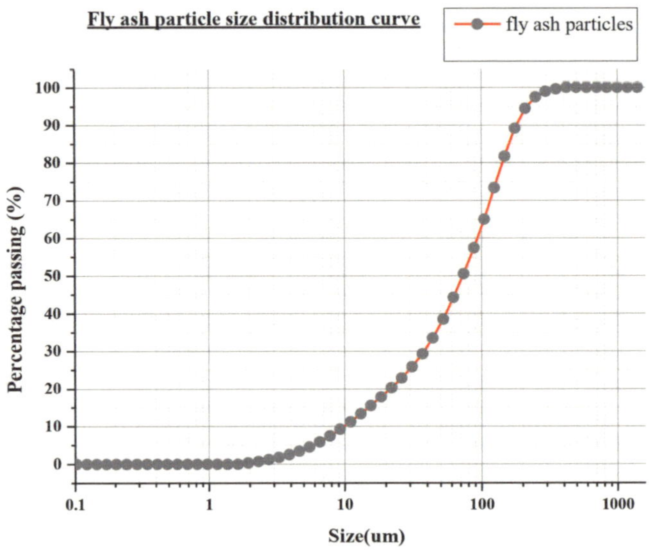

Figure 6.
Particle size distribution for fly ash, used in the experiment.

Mix identity	Cement	Sand	Flyash	Bottomash	Stone Aggregate	Water-cement ratio
CC	1	1.60	—	—	2.40	0.50
BC-1	1	1.44	—	0.16	2.40	0.50
BC-2	1	1.28	—	0.32	2.40	0.50
BC-3	1	1.12	—	0.48	2.40	0.50
BC-4	1	0.96	—	0.64	2.40	0.50
BC-5	1	0.80	—	0.80	2.40	0.50
BC-6	1	0.64	—	0.96	2.40	0.50
BC-7	1	0.48	—	1.12	2.40	0.50
BC-8	1	0.32	—	1.28	2.40	0.50
BC-9	1	0.16	—	1.44	2.40	0.50
BC-10	1	0.00	—	1.60	2.40	0.50
FC-1	1	1.44	0.16	—	2.40	0.50
FC-2	1	1.28	0.32	—	2.40	0.50
FC-3	1	1.12	0.48	—	2.40	0.50
FC-4	1	0.96	0.64	—	2.40	0.50
FC-5	1	0.80	0.80	—	2.40	0.50
FC-6	1	0.64	0.96	—	2.40	0.50
FC-7	1	0.48	1.12	—	2.40	0.50
FC-8	1	0.32	1.28	—	2.40	0.50
FC-9	1	0.16	1.44	—	2.40	0.50
FC-10	1	0.00	1.60	—	2.40	0.50

Table 2.
Design mix concrete of M-25 grade with fly ash/bottom ash substitution separately.

Sample identity	Cement	Lime dust	Fly ash	Bottom ash	Stone aggregate	Water-cement ratio
42(75:25)	1.00	0.40	—	1.20	2.40	0.50
41(67:33)	1.00	0.53	—	1.07	2.40	0.50
40(50:50)	1.00	0.80	—	0.80	2.40	0.50
39(75:25)	1.00	0.40	1.20	—	2.40	0.50
38(67:33)	1.00	0.53	1.07	—	2.40	0.50
37(50:50)	1.00	0.80	0.80	—	2.40	0.50

Table 3.
Design mix concrete of M-25 grade lime dust and fly ash/bottom ash (no sand).

Mortar mix identity	Mortar mix (MM3 grade)	Cement wt. ratio	Sand wt. ratio	Flyash wt. ratio	Bottomash wt. ratio
Control	1 cement:6 and	1.00	6.00	—	—
A-1	1 cement:(5.4 sand + 0.6 flyash)	1.00	5.40	0.60	—
A-2	1 cement:(4.8 sand + 1.2 flyash)	1.00	4.80	1.20	—
A-3	1 cement:(4.2 sand + 1.8 flyash)	1.00	4.20	1.80	—
A-4	1 cement:(3.6 sand + 2.4 flyash)	1.00	3.60	2.40	—
A-5	1 cement:(3.0 sand + 3.0 flyash)	1.00	3.00	3.00	—
A-6	1 cement:(2.4 sand + 3.6 flyash)	1.00	2.40	3.60	—
A-7	1 cement:(1.8 sand + 4.2 flyash)	1.00	1.80	4.20	—
A-8	1 cement:(1.2 sand + 4.8 flyash)	1.00	1.20	4.80	—
A-9	1 cement:(0.6 sand + 5.4 flyash)	1.00	0.60	5.40	—
A-10	1 cement:6 flyash	1.00	0.00	6.00	—
B-1	1 cement:(5.4 sand + 0.6 bottomash)	1.00	5.40	—	0.60
B-2	1 cement:(4.8 sand + 1.2 bottomash)	1.00	4.80	—	1.20
B-3	1 cement:(4.2 sand + 1.8 bottomash)	1.00	4.20	—	1.80
B-4	1 cement:(3.6 sand + 2.4 bottomash)	1.00	3.60	—	2.40
B-5	1 cement:(3.0 sand + 3.0 bottomash)	1.00	3.00	—	3.00

Mortar mix identity	Mortar mix (MM3 grade)	Cement wt. ratio	Sand wt. ratio	Flyash wt. ratio	Bottomash wt. ratio
B-6	1 cement:(2.4 sand + 3.6 bottomash)	1.00	2.40	—	3.60
B-7	1 cement:(1.8 sand + 4.2 bottomash)	1.00	1.80	—	4.20
B-8	1 cement:(1.2 sand + 4.8 bottomash)	1.00	1.20	—	4.80
B-9	1 cement:(0.6 sand + 5.4 bottom ash)	1.00	0.60	—	5.40
B-10	1 cement:6 bottomash	1.00	0.00	—	6.00

Table 4.
Masonry mortar mix of MM 3 grade with fly ash/bottom ash substitution separately.

Mortar mix identityy	Mortar Mix (MM3 grade)	Cement	Sand	Fly ash	Bottom ash	Lime dust	Marble dust
1	1 cement:6 sand	1.00	6.00	—	—	—	—
2	1 cement:6 (3.0 limedust+3.0 flyash)	1.00	—	3.00	—	3.00	—
3	1 cement:6 (3.0 marbledust + 3.0 flyash)	1.00	—	3.00	—	—	3.00
4	1 cement:6 flyash	1.00	—	6.00	—	—	—
5	1 cement:6 (3.0 limedust + 3.0 bottomash)	1.00	—	—	3.00	3.00	—
6	1 cement:6 (3.0 marbledust + 3.0 bottomash)	1.00	—	—	3.00	—	3.00
7	1 cement:6 bottomash)	1.00	—	—	6.00	—	—

Table 5.
Masonry mortar mix of MM 3 grade with fly ash/bottom ash and lime dust/marble dust.

Two wooden frames having dimension 500 × 500 were built in order to construct and hold the brick wall samples within those. This was done mainly in order to make it a modular system which could be portable enough so that after experimentation further developments of the samples like plastering, coloring etc. are feasible. Clay bricks, Portland Pozzolana Cement (PPC), sand, fly ash, lime dust and water were the raw materials used for the construction of the brick walls. Dimension of standard Indian burnt clay bricks are 230 × 115 × 75. Two sets of brick walls were developed for the experimentation purpose, one with conventional plaster grade MM5 with cement-sand ratio 1:4, another one with same grade of mortar and plaster but with 50% Lime dust and 50% fly ash in place of 100% sand. Dimensions of the both brick walls were 480 × 480 × 115 and 480 × 480 × 115 with 12 mm plaster on either side.

Figure 7.
Guarded hot box test setup with data logging and wall panel under preparation.

4. Experimental results

4.1 Compressive strength and thermal conductivity test results

As described in previous chapter, i.e. Chapter 3, all the design and nominal mix samples in respect of Concrete of various grades starting from M-15 to M-25 were put to test to determine various physical and thermal parameters. Similarly, Mortar mix samples of various proportions with respect to two most used grades MM3 and MM5 were put to tests. The tests performed on those samples were of destructive and non-destructive in nature. The tests are essential for durability and application worthiness of such concrete and mortar mixes (**Figures 8–15**).

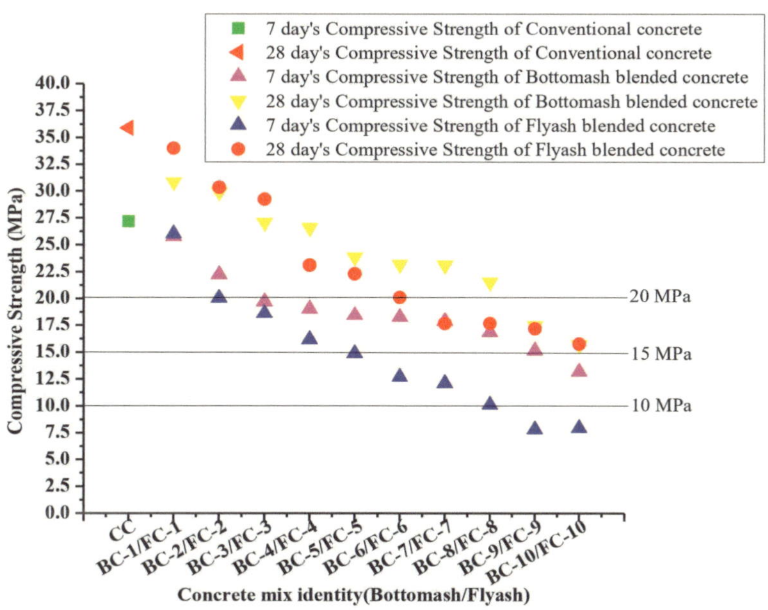

Figure 8.
*Compressive strength of concrete mix (refer **Table 2** for mix proportion).*

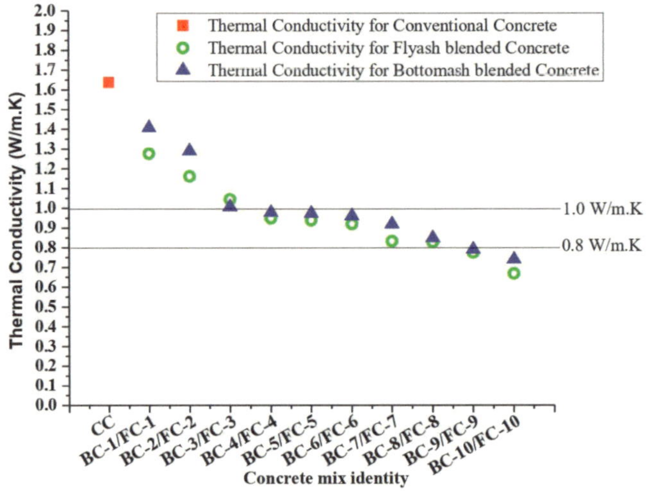

Figure 9.
*Thermal conductivity test results of concrete mix (refer **Table 2** for mix proportion).*

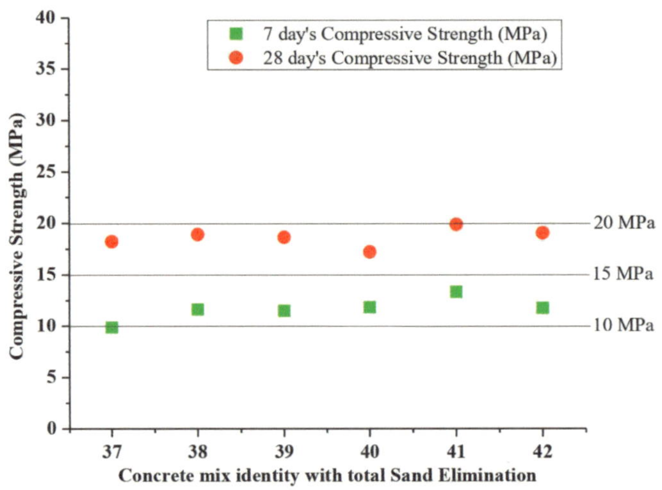

Figure 10.
*Compressive strength test of concrete mix (refer **Table 3** for mix proportion).*

It may be seen from the above plotted results, bottomash blended concrete offers M-25 grade strength up to 40% replacement and M-20 grade strength up to 80% replacement of sand. Fly ash blended concrete gives marginally lower results.

It may be seen from the above plotted thermal conductivity test results, both fly ash and bottom ash blended concrete offer reduced thermal conductivity than conventional concrete mix.

From the above plotted result, it may be observed that flyash-lime dust (38) and bottomash-limedust (41) blended mix in the ratio of 67:33 offer M-20 grade strength and same combination with 75:25 ratio offer close to M-20 grade strength.

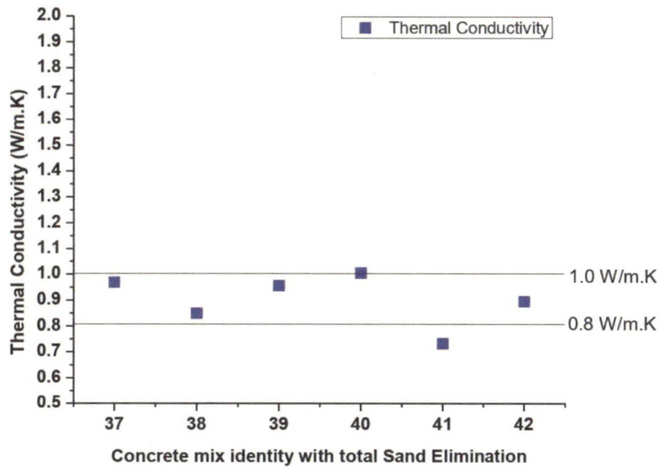

Figure 11.
Thermal conductivity test of concrete mix (refer ***Table** 3 for mix proportion).*

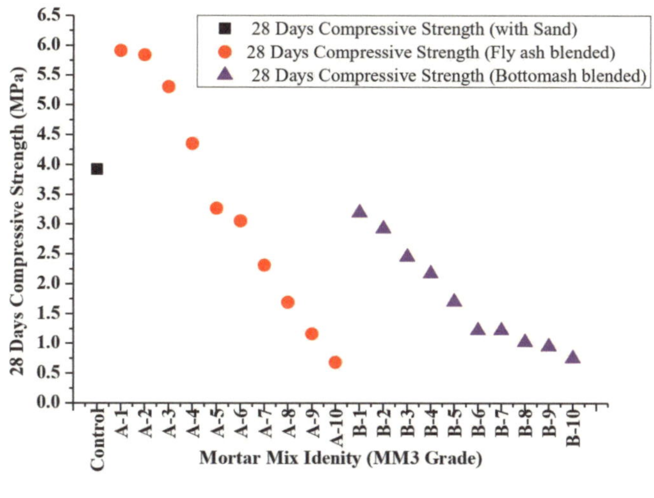

Figure 12.
Compressive strength test of mortar mix (MM3) (refer ***Table** 4 for mix proportion).*

Mixes 38 (flyash:limedust::67:33) and 41 (bottomash:limedust::67:33) offer lower thermal conductivity value.

From the above plotted result, it may be observed that up to 60% replacement of sand by flyash, strength remains within MM-3 grade required criteria.

Reduction in thermal conductivity value follows same trend by fly ash and bottom ash blended mortar mixes respectively.

Flyash-limedust (2) and bottomash-limedust combination offer MM-3 grade strength compatibility.

Considerable reduction in thermal conductivity values observed in both flyash (2, 3, 4) and bottomash (5, 6, 7) blended mixes with respect to the conventional mix (1).

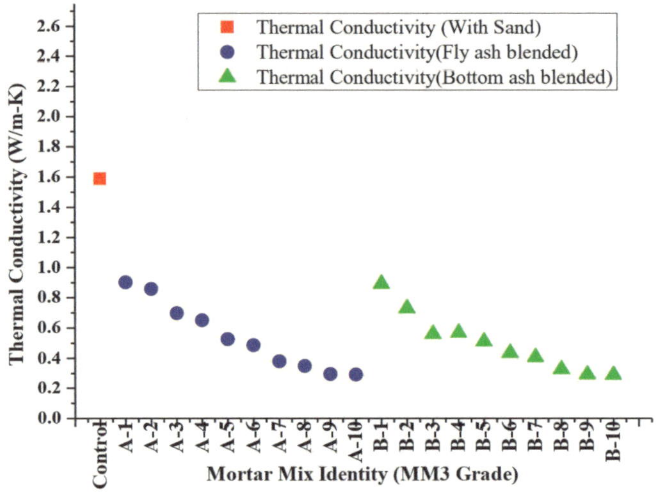

Figure 13.
Thermal conductivity test results of mortar mix (refer ***Table 4*** *for mix proportion).*

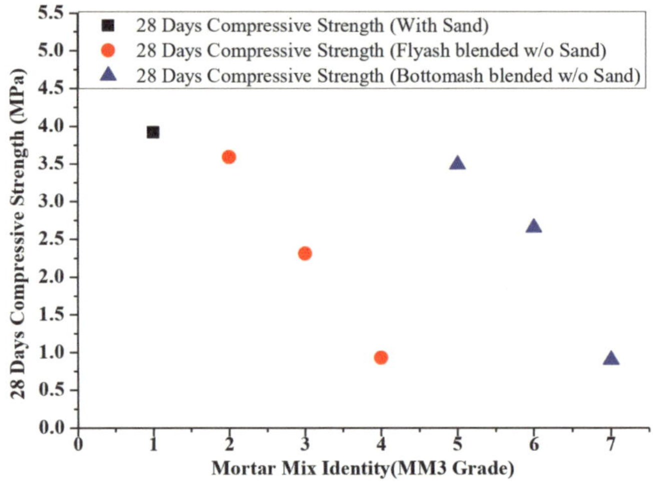

Figure 14.
Compressive strength test values of MM3 grade mortar (total replacement of sand by flyash, lime dust, marble dust, bottomash) (refer ***Table 5****).*

4.2 Test result for overall heat transfer co-efficient (U-value)

The hot side temperature of 40°C, and cold side temperature of 25°C were maintained for 3 consecutive days. Both the brick wall panels with conventional mortar and plaster combination and fly ash-lime dust combination were tested under identical test parameters. In steady state condition, average standard deviation in brick surface temperature of both days of testing was 0.056°C on both the hot and cold side. The experimentally obtained U-values for both the cases and final difference thereof is shown in **Table 6**.

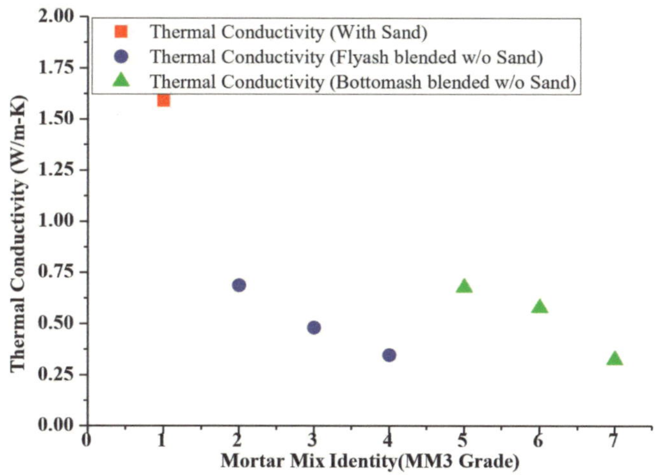

Figure 15.
*Thermal conductivity test results of mortar mix (refer **Table** 5).*

Wall sample with cement-flyash-lime mortar mix		Wall sample with cement-sand mortar mix	
Days	**U-value (W/m^2 K)**	**Days**	**U-value (W/m^2 K)**
Day 1	3.051	Day 4	3.655
Day 2	2.954	Day 5	3.534
Day 3	3.050	Day 6	3.536
Average of 3 days	3.018	Average of 3 days	3.575
Difference in U-value (%)		15.58	

Table 6.
Final result of wall panel U-value.

5. Economical and environmental benefits

A square shaped room of size 10 m^2 plan area and four wall size each of 10 m^2 is considered for such energy and economic analysis at Kolkata, India location.

5.1 Assumptions

i. The outside average temperature in a month is considered as 30°C, and inside conditioned temperature as 25°C during entire duration of activity of 8 h in a day.

ii. Active days in a month is considered as 26 days.

iii. Average electricity tariff considered as Rs.8/- per unit of electricity consumed.

Room (m^2)	Walls (m^2)	Diff. in temp (°C)	U-value for roof assembly (W/ m^2 K)	U-value for wall assembly (W/ m^2 K)	Active hours in a month (h)	Total energy flow thro' roof (kWh)	Total energy flow thro' walls (kWh)	Gross energy (kWh)	Savings per month (Rs.)	CO_2 saved per month (kg)
10	40	5	3.3371	3.5750	208	34.71	148.72	183.43		
10	40	5	2.6187	3.0180	208	27.23	125.55	152.78	245.20	25.13

Table 7.
Economic and environmental analysis.

iv. RCC (of M-20 grade blended concrete) roof slab thickness considered as 125 mm, overlaid by 50 mm thick PCC (of M-15 grade blended concrete), and topping by 25 mm thick PCC (M-15 grade blended concrete without sand).

v. Masonry wall thickness 125 mm, made of burnt clay brick and mortar (of MM-3 grade), plastered both sides with 12 mm thick plaster of same grade.

vi. CO_2 emission due to energy generation from thermal power plant is considered as 0.82 kg per kWh

6. Conclusions

From the study pertaining to this work, the following can be concluded:

1. Fly ash and Bottom ash, the 100% utilization of which is still not possible, might follow a positive note. Building industry could be immensely benefitted by such usage from the perspectives of 3Es, Economy, Environment and Energy. The benefit for the case example under Section 5 is shown in **Table 7.**

2. Other than electrical energy saving due to the inherent insulating nature, coal ash is available free of cost from all thermal power plants and even transportation cost up to 100 km radius is reimbursed in India by the Plant authorities. The mechanical dredging required to extract sand from river bed is totally avoided in such substitution.

3. Since the rate of formation outpace rate of extraction due to infrastructure developmental need, sand is being depleted rapidly worldwide. Rampant sand mining from river bed is causing serious environmental imbalances like lowering of water table in the adjacent agricultural field, river bank erosion, disturbing effect on aquatic life etc. At the same time, continuous accumulation of coal combustion residues since ages, pose serious health threats in the adjoining areas of ash dykes. These two issues could be compensated by the proposed substitution.

4. The research work did not involve any energy consuming machinery or technique and no synthetic additive was used.

5. Energy saving out of such substitution is one of the major finding, which would contribute to abatement of rising CO_2 emission. Substitution in masonry mortar

could lower U-value by 15.58% (**Table 7**), which is translated into reduction in considerable electrical energy requirement to maintain comfort condition in buildings.

6. Such substitution of coal ash in concrete and mortar mixes in building construction could also contribute significantly in the concept of Near Zero Energy Building by restricting its specific energy demand up to certain extent. It could also provide impetus to Green and Affordable Housing program.

Acknowledgements

The work was integral part of Doctoral research study at Jadavpur University and at CSIR-Central Glass & Ceramic Research Institute without any financial grant. Prof. Indranil Manna, former Director of CSIR-Central Glass & Ceramic Research Institute had kindly permitted to pursue the doctoral research work, and Prof. Subhasis Neogi and Dr. Arup Ghosh had acted as Supervisor and Co-Supervisor for the work. Mr. Debasish Chowdhury had helped in carrying out U-value tests and Mr. Subrata Mistry had co-operated tirelessly in preparing around 200 samples. Mr. G. Debnath, Mrs. M. Kanu, Mrs. Swati Samanta, Mr. Nitai Dey, Mr. Ashok Mndal of CSIR-CGCRI had helped in material characterization and Mr. Bidhan Choudhury and Dr. Chandana Patra of CSIR-CGCRI had helped immensely in collection of some reference articles.

Conflict of interest

The author declare that there remains no scope for conflict of interest.

Author details

Avijit Ghosh
CSIR—Central Glass and Ceramic Research Institute, Kolkata, India

*Address all correspondence to: avijitenergy@gmail.com

References

[1] World Energy Outlook 2021. International Energy Agency. Available from: https://www.iea.org/reports/world-energy-outlook-2021/executive-summary

[2] Shahzad Baig K, Yousaf M. Coal fired power plants:emissions problems and controlling techniques. Journal of Earth Science and Climatic Change. 2017;**8**:404

[3] 2021 Global Status Report for Buildings and Construction. Global Alliance for Buildings and Construction. UNEP and IEA. Available from: https://globalabc.org/sites/default/files/2021-10/GABC_Buildings-GSR-2021_BOOK.pdf

[4] Tracking Buildings 2021. International Energy Agency. Available from: https://www.iea.org/reports/tracking-buildings-2021

[5] Ramesh T, Ravi P, Shukla KK. Life cycle energy analysis of buildings: An overview. Energy and Buildings. 2010;**42**:1592-1600

[6] International Energy Agency (IEA). Transition to Sustainable Buildings—Strategies and Opportunities to 2050. Paris; 2013

[7] Tae S, Kim T, Chae UC. Analysis of environmental impact for concrete using LCA by varying the recycling components, the compressive strength and the admixture material mixing. Sustainability. 2016;**8**(389):1-14. DOI: 10.3390/su8040389

[8] Harvey DLD. Net climatic impact of solid foam insulation produced with halocarbon and non-halocarbon blowing agents. Building and Environment. 2007;**42**:2860-2879

[9] Bergey D. Net Global warming due to insulating foams, Northeast sustainable energy association conference. Building Science Corporation; March 10. 2010

[10] Gupta NK, Sharma AK, Sharma A. Quantifying embodied energy using green building technologies under affordable housing construction. Open Journal of Energy Efficiency. 2013;**2**: 171-175. DOI: 10.4236/ojee.2013.24022

[11] IS 2250. Indian Standard Code of Practice for Preparation and Use of Masonry Mortars (First Revision), Manak Bhavan, 9, Bahadur Shah Zafar Marg, New Delhi 110002: Bureau of Indian Standards; 1981 (Reaffirmed in 2000)

[12] Higgins D. Sustainable concrete: How can additions contribute. In: Annual Technical Symposium; March 28. UK: The Institute of Concrete Technology; 2006

[13] Kumar D, Gupta A, Ram S. Use of bottom ash in the replacement of fine aggregate for making concrete. International Journal of Current Engineering and Technology. 2014;**4**(6): 3891-3895

[14] Kadam MP, Patil YD. Effect of coal bottom ash as sand replacement on the properties of concrete with different w/c ratio. International Journal of Advanced Technology in Civil Engineering. 2013;**2**(1):45-50

[15] Kadam MP, Patil YD. American Journal of Civil Engineering and Architecture. 2014;**2**(5):160-166. DOI: 10.12691/ajcea-2-5-2

[16] Cadersa AS, Auckburally I. Use of unprocessed coal bottom ash as partial fine aggregate replacement in concrete.

University of Mauritius Research Journal. 2014;**20**:62-84

[17] Aggarwal P, Aggarwal Y, Gupta SM. Effect of bottom ash as replacement of fine aggregates in concrete. Asian Journal of Civil Engineering (Building and Housing). 2007;**8**(1):49-62

[18] Kumar G, Dinesh M, Younus M, Haleem M. An experimental study on partial replacement of fine aggregate with coal bottom ash in concrete. International Journal of Research Sciences and Advanced Engineering. 2016;**2**(15):39-49

[19] Abbas M, Kumar R, Kumar M. Study the effect of coal bottom ash and limestone dust as partial replacement of sand and cement. International Journal of Scientific Research and Education. 2016;**4**(5):5363-5372. DOI: 10.18535/ijsre/v4i05.11

[20] Rafieizonooz M, Mirza J, Salim MR, Hussin MW, Khankhaje E. Investigation of coal bottom ash and fly ash in concrete as replacement for sand and cement. Construction and Building Materials. 2016;**116**:15-24

[21] Rafieizonooz M, Salim RM, Mirza J, Hussin MW, Salmiati KR, Khankhaje E. Toxicity characteristics and durability of concrete containing coal ash as substitute for cement and river sand. Construction and Building Materials. 2017;**143**:234-246

[22] Gencel O, Bilir T, Bekir I T. Properties of mortars with flyash as fine aggregate. Construction and Building Materials. 2015;**93**:782-789

[23] Brake NA, Oruji S, Nalluri L, Guduru RK. Strength activity and microstructure of blended ultra-fine coal bottom. Construction and Building Materials. 2017;**153**:317-326

[24] Kim KH. Utilization of sieved and ground coal bottom ash powders as a coarse binder in high strength mortar to improve workability. Construction and Building Materials. 2015;**91**:57-64

[25] Shahidan S, Ramzi NIR, Maarof MZ, Noorwirdawati A. Physical and chemical properties of coal bottom ash from Tanjung Bin Power Plant, International Engineering Research and Innovation Symposium (IRIS), IOP Conference Series. Materials Science and Engineering. 2016;**160**:012056

[26] Abbas et al. Study the effect of coal bottom ash and lime stone dust as partial replacement of sand and cement. International Journal of Scientific Research And Education. 2016;**4**(5):5363-5372

[27] Ghosh Avijit, Ghosh Arup, Neogi S. Evaluation of physical and thermal properties of coal combustion residue blended concrete for energy efficient building application in India. Advances in Building Energy Research. 2018:1-23. DOI: 10.1080/17512549.2018.1557076

[28] Ghosh Avijit, Ghosh Arup, Neogi S. Reuse of fly ash and bottom ash in mortars with improved thermal conductivity performance for buildings. Heliyon. 2018;**4**(2018):e00934. DOI: 10.1016/j.heliyon.2018.e00934

[29] IS 383. Indian Standard Code of Practice for Coarse and Fine Aggregate for Concrete (Third revision). Manak Bhavan, 9, Bahadur Shah Zafar Marg, New Delhi 110002: Bureau of Indian Standards; 2016

[30] IS 456. Indian Standard Code of Practice for Plain and Reinforced Concrete (Fourth Revision). Manak Bhavan, 9, Bahadur Shah Zafar Marg, New Delhi 110002: Bureau of Indian Standards; 2000 (Reaffirmed in 2005)

[31] IS 10262. Indian Standard Code of Practice for Concrete Mix Proportioning

Guidelines (Second Revision). Manak Bhavan, 9, Bahadur Shah Zafar Marg, New Delhi 110002: Bureau of Indian Standards; 2019

[32] SP 23. Handbook on Concrete Mixes. Manak Bhavan, 9, Bahadur Shah Zafar Marg, New Delhi 110002: Bureau of Indian Standards; 1982

Section 2

Design

Chapter 4

Wooden Facade Renovation and Additional Floor Construction for Suburban Development in Finland

Markku Karjalainen, Hüseyin Emre Ilgın, Lauri Metsäranta and Markku Norvasuo

Abstract

Finnish urban settlements are in the age of restoration, and the suburbs need improvements in Finland. In this sense, wooden facade renovation and additional floor construction are viable and sustainable solutions for this development in the Finnish context. This chapter focuses on these important applications from the Finnish residents' perspective as ecologically sound engineering solutions through a survey. In doing so, the challenges of facade renovation, as well as the benefits of additional floor construction, were presented. The main purpose of the survey was to get the opinions of the residents, find out which variables are important, make inferences for the planning and improvement of such areas, and determine what will be emphasized in the sustainable suburban development of the future. Therefore, the results were based on this empirical approach—survey—but further research such as energy analysis, wood-based facade renovation, and additional floor solutions will be done as part of other studies. It is believed that this study will contribute to the use of sustainable materials and decarbonization of buildings as well as zero energy building (nZEB) to overcome the challenges posed by climate change by the diffusion of wood in the renovation of buildings.

Keywords: timber/wood, facade renovation, additional floor construction, suburban development, zero energy building (nZEB), sustainability, Finland

1. Introduction

Over 220 million building units built before 2001 accounted for 85% of the EU's building stock, of which about 90% will continue to be in service in 2050 [1–3]. However, most are not energy-efficient and use fossil fuels and older technologies and about 40% of the EU's total energy consumption, and more than 35% of energy-related greenhouse gas emissions originate from buildings [4]. In this sense, to achieve a 55% emission reduction in the 2030 Climate Goal Plan, European countries must make a significant 60% reduction in the greenhouse gas emissions of buildings [5].

The abovementioned scenario was highlighted in the Renovation Wave, which needed changes in current construction and renovation practices in the industry and supported a combination of strong efficiency measures alongside the phasing out of fossil fuels and the transition to renewable energy [6]. Due to the aging of the European building stock, building owners and the entire building sector are faced with extensive renovation works. The renovation and refurbishment of the building facades and external walls are among the most critical tasks to be undertaken [7].

Therefore, it becomes urgent to focus on refurbishing existing building stock within the principles of ecologically responsive engineering to make it more energy-efficient and less carbon-intensive. In addition to reducing energy bills and emissions, renovation can provide many possibilities with social, environmental, and economic benefits such as making buildings more durable, healthier, greener, more accessible, and smarter. However, the current deep renovation rate of 0.2% needs to grow by at least 10 to 2% and approach 3% as quickly as possible [8].

Here, it is worth mentioning that "Net Zero Energy Buildings" (nZEBs) will be the next major frontier for innovation and competition in the world real estate market and can scale rapidly in Europe [9]. In this sense, European energy policies set the nZEB target [10] to promote the energy transition of the construction sector. EU programs, notably "Horizon 2020," introduce the nZEB design as well as its evolution to the Positive Energy Generation (PEB) model [11].

Moreover, Scandinavian countries are working toward regional carbon neutrality before the goals of the European Union. Finland aims for carbon neutrality by 2035 and is developing several policies, including low-carbon construction legislation [12]. The new approach includes normative carbon limits for different building types before 2025.

Similar to the aforementioned EU building stock situation, which is poor in terms of energy efficiency, most of the building stock in Finland was built between the 1970s and 90s and needs serious renovation, with residential buildings accounting for 63% of the total gross floor area [13, 14]. Nearly a third of the housing stock that makes up a significant portion of the Finnish building stock was poorly insulated

Figure 1.
A typical suburban apartment in Finland.

DOI: http://dx.doi.org/10.5772/intechopen.101620

suburban apartments from the 1960s and 1970s and in need of refurbishment [15, 16] (see **Figure 1**).

It is worth noting here of the total annual construction expenditures in Finland, approximately 47% is spent on infrastructure projects, 21% on new buildings, and 32% on renovations [17], in which low energy efficiency, lack of balcony, lack of elevator, and unpleasant appearance are among the critical issues identified for Finnish suburban apartments [18]. On the other hand, in practice, apartment renovation is a slow and expensive process that requires a lot of capital and government subsidies [19].

At this point, wooden additional floor construction (**Figure 2**) stands out as an ecological engineering solution with many advantages such as being environmentally friendly, providing a significant increase in the gross floor area and energy efficiency of the existing building, and improving esthetic appearance [20].

Furthermore, additional floors essentially increase the building's energy efficiency directly, but also indirectly, for example, by using the revenues from the building for energy regeneration. These energy-related measures help increase the cost-effectiveness of the entire renovation process and maintenance of buildings [21–23]. Additional floors do not significantly increase the overall energy consumption of the upgraded building, and as passive energy-efficient structures, they can significantly increase the energy efficiency of refurbished buildings, especially if the upper floors have not been renovated for a long time [24].

Similarly, one of the most effective energy-saving measures for buildings is facade renovation and roof/attic insulation with wood-based solutions in the building envelope [25–30]. The amount of this saving varies according to the system and material used such as 50 mm thermal-bridge-breaking on-site mounted additional isolation (total U-value 0.26 W/m^2K), a modular prefabricated facade renovation system (total U-value 0.18 W/m^2K) [31]. Moreover, some prefabricated and integrated facade

Figure 2.
Representative image of wooden additional floor construction.

module solutions offer the possibility to improve the current energy performance up to zero energy, while ensuring minimum disturbance for the occupants, during and after the renovation [32, 33].

As in many other projects, material selection has critical importance in renovation projects such as facade renovation (**Figure 3**) and therefore additional floor construction. Wood as a renewable material is ecological and environmentally friendly: One cubic meter of growing wood can hold about one ton of CO_2 from the atmosphere, the mass of wood is about 500 kg/m^3, and half of this mass is carbon = 250 kg/m^3. One of our best allies in solving the climate crisis due to its potential eco-friendly properties, wood is at the forefront of tackling European climate policy [34–37]. Furthermore, due to its significantly lower carbon footprint and potential cost-effectiveness compared with conventional materials such as reinforced concrete and steel, and numerous positive effects on the environment combined with technological advances [38–41]. Besides this, as it is well known, from an architectural point of view, wooden buildings are thought to have the potential to generate a more pleasant, warm, and natural environment.

Thus, renovating and expanding existing buildings with wood can contribute significantly to sustainable urban redevelopment. Renovation of building envelopes (e.g., roof, facade) with highly insulated wooden components can significantly reduce the conduction heat losses of existing buildings and the associated heating energy demand [42]. In addition, the characteristics (e.g., load-bearing capacity, flat roof) of Finnish suburban apartment blocks from the 1960s and 1970s and the current Finnish fire code allow light additional floor construction.

In literature, there is a limited number of research on residents' or consumers' attitudes toward the use of timber in building construction [43]. Important research over the past 10 years has reported perceived benefits and barriers to consumers' use of wood as a building material. Among these studies, Lähtinen et al. scrutinized the ecological, physiological-technological, esthetic, and welfare properties of wood as a building material from the Finnish perspective [44]. Environmental features and

Figure 3.
Representative image of wooden facade renovation.

esthetics concerns were assessed as the most important advantage of wood in several studies [45–47], while coziness and longevity were highlighted as other benefits [41, 44]. On the other hand, some studies [46–48] showed that users are skeptical of the use of wood as a structural system material on certain issues. Such as durability, maintenance, structural performance, and fire safety. In addition, there are few studies on the construction of wooden additional floors, among which, Karjalainen et al. [20] focused on the various stages and benefits of wooden additional floor construction for the Finnish housing and real estate companies, and Soikkeli [49] highlighted the financial and practical advantages of developing an industrial scale model for wooden additional floor construction.

No study has been found on the perceptions of the residents regarding the renovation of wooden facades and the construction of additional floors in literature. At this point, it is worth noting that the acceptability of a new construction method by users or residents is important to ensure its sustainability and diffusion as a contributor to the Finnish forest-based bioeconomy. It is believed that the study will make an important contribution to this issue. This chapter focuses on wooden facade renovation and additional floor construction through a resident survey as ecologically sound engineering solutions to contribute to the decarbonization of buildings and a zero-energy building approach. In doing so, the challenges of facade renovation and the benefits of additional floor construction are presented.

In this study, timber or wood refers to engineered wood products such as cross-laminated timber [(CLT) a prefabricated multi-layer EWP, manufactured from at least three layers of boards by gluing their surfaces together with an adhesive under pressure], laminated veneer lumber [(LVL) made by bonding together thin vertical softwood veneers with their grain parallel to the longitudinal axis of the section, under heat and pressure)], and glue-laminated timber (glulam) [(GL) made by gluing together several graded timber laminations with their grain parallel to the longitudinal axis of the section)].

2. Wooden facade renovation and additional floor construction

Facade renovation has many advantages (e.g., esthetic improvement, energy-saving, increased thermal comfort, reducing CO_2 emissions, and improving the quality of the built environment) as demonstrated in many EU projects [50–54]. The most common facade renovation technologies and applications consist of installing external and internal insulation, enhancing airtightness, installation of photovoltaic panels, heat recovery, and installation of efficient heating, ventilation, and air conditioning (HVAC) systems [55].

It is worth mentioning here the main barriers and challenges encountered particularly in deep renovation projects as follows (**Table 1**) [56, 57].

On the other hand, the advantages of additional floor construction that contribute to overcoming the abovementioned obstacle can be summarized as follows [20]: (i) promote beneficial development of the building stock and increase property owners' incomes; (ii) provide short-term income to housing companies by selling additional floors and proceeds to be used to finance the renovation of existing property, such as the renovation of an elevator to improve the building's accessibility and commercial conditions; (iii) although it significantly increases the total floor area, it does not significantly increase the overall energy consumption of the upgraded building. (iv) significantly improve the energy efficiency of older buildings as passive energy-efficient

Barrier typology	Barrier
Embedded market inefficiencies	split incentives and conflicts of interest between the building owner and tenants
Informative-social	lack of information dissemination and convincing end users of the benefits of deep renovations
	time-consuming and complex decision-making processes
	lack of consensus and support from residents, which generally hinders effective approval of interventions
	inconvenience during site, studies and relocation of users
	lack of communication between different interested parties
	low awareness of energy efficiency and non-energy benefits of refurbishment
Financial	limited financing options offered and limited third-party financier involvement
	lack of satisfactory financial support, especially for low-income homeowners
	limited impact of Energy Performance Certificate i on property value
	lack of trust in investors
	long payback periods
	limited financing/ insufficient budgets - high up-front costs and owners' reluctance to borrow for energy replacement purposes
Organization and structure of the renovation market	Difficulties in coordinating communications with other relevant stakeholders
	Insufficient resources on part of small and medium-sized enterprises for larger tenders
Regulatory, Knowledge informative based technical	lack of continuity in regulations
	limited government subsidies and programs in specific regions
	lack of skills and lack of training
Technical	performance gaps and uncertainty
	lack of reliable and standardized or integrated solutions to meet the various building standards requirements related to energy savings
	inadequacies in technical solutions
	safety risk associated with deep renovation processes
	users' lack of technical knowledge and confidence in the effectiveness of savings in energy regeneration

Table 1.
Main barriers and challenges of the (deep) renovation process.

structures, especially if the upper floors have not been renovated for a long time; (v) improve the image and appearance of the building; and (vi) advantageous in terms of carbon footprint over demolition and new construction.

3. Research methods

This study was carried out mainly as a literature review including international peer-reviewed journals and similar research projects, supported by materials collected during "Effects of wooden buildings on neighborhood-level" at the Tampere University—a project that is part of the Ministry of the Environment's Growth and

Development from Wood support program involves cross-sectional data from the Pukinmäki-Savela (postcode 00720) area in the City of Helsinki, Finland [58]. The focus of the project was to study the attitudes of residents and users of a neighborhood toward wooden buildings and to use this information in the planning and infill construction of urban areas. One potential method is the construction of additional floors to suburban apartment houses, as an ecologically sensitive engineering solution to support the decarbonization of buildings and a zero-energy building approach.

It is worth mentioning here that Pukinmäki is a district in the northeastern part of Helsinki. The area was added to Helsinki in 1946, and the first few apartment blocks were built in the 1960s. Most of the apartments are from the 1970s and 1980s. On the other hand, Savela (**Figure 4**) is a residential area in the Pukinmäki district. In the 1960s and early 1970s, Savela was under threat of partial rezoning and segregation of single-family homes in the city as green space. However, over time, the area turned from a detached area to an apartment area, with mostly low-rise apartments built in the 1980s and 1990s.

The sample, 800 Finnish-speaking people aged 18–69 in the area, was chosen randomly. A total of 243 responses were received for the entire survey, corresponding to a response rate of 30%. The survey data generally represented the population of the selected area, but there were also minor differences in representation among background variables such as gender and age. For example, in terms of gender, 54 and 57% were female, 46 and 43% were male in the whole population and sample, respectively. On the other hand, in terms of age groups, the older population was slightly overrepresented, while younger respondents were slightly underrepresented; in the whole population and sample, 20 and 30% were aged 60–69, and 25 and 15% were aged 18–29, respectively. As regards education level, high school and university graduates constituted 73% of the entire population; this rate was 90% in the sample group.

Focusing on facade renovation and additional floor construction, the survey was divided into five main parts. The first part was about background information, and in the second part, the participants were asked about their opinions of timber in buildings. The third part was about the dwelling preferences (number of floors, facade material, etc.), and in the fourth part, the opinions on the wooden Eskolantie apartment houses were asked. In the last part, six different renovation alternatives were

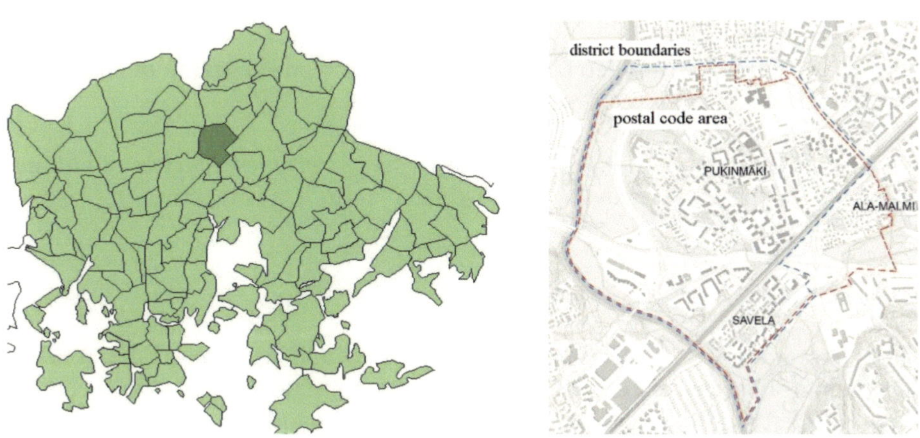

Figure 4.
Pukinmäki—Savela area in the city of Helsinki (Finland).

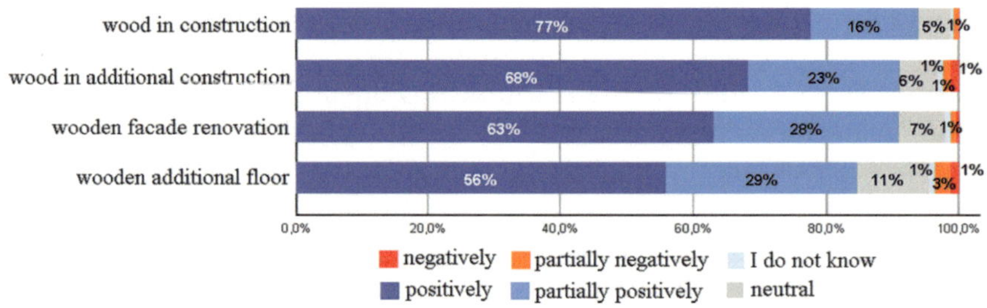

Figure 5.
Residents' perception of wood in (additional) construction, wooden facade renovation, and wooden additional floor.

presented, and the residents' opinions on wooden additional floor construction and facade renovation were asked.

4. Main findings from suburban residents' survey

The survey results mainly highlighted the following (**Figure 5**):

- respondents supported suburban development through wood refurbishment and additional construction.
- attitudes toward wood in construction were positive, and wood was perceived as an ecological alternative in construction.
- residents of suburban settlements preferred low-rise and low-dense suburban fabric, and wood was considered the most pleasant facade material.
- wooden structures were also generally perceived as more beautiful, more ecological, and healthier than buildings made of other materials.
- concerns about wood material were mainly related to its technical properties, such as fire safety, long-term durability, and maintenance needs of facades.
- it was reported that wooden facade renovation and additional floor construction can increase the attractiveness of residential areas.
- apartment owners would welcome their housing company's decision to implement wooden facade renovation and additional floor construction.

5. Conclusion

This chapter focuses on wooden facade renovation and additional floor construction as ecologically sensitive engineering solutions from the Finnish residents' point of view. In doing so, the challenges of facade renovation and the benefits of additional floor construction are presented.

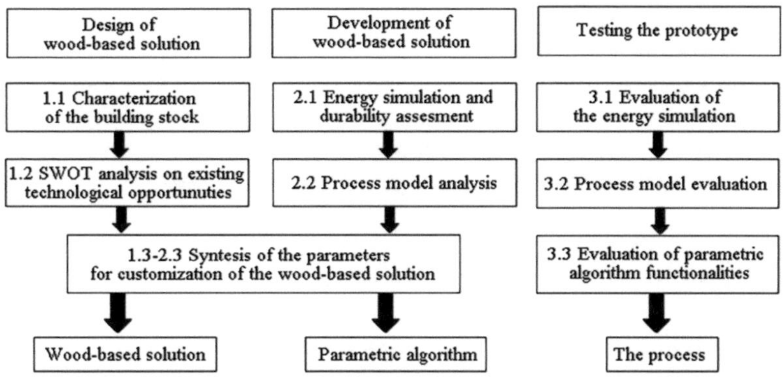

Figure 6.
Suggested methodology for future studies.

In the context of the transition to service-oriented business solutions in Europe, future and emerging technologies in facade renovation and additional floor construction can make a difference with the risks of overheating from climate change and the reduction in energy demand for cooling in existing buildings, considering the human-centered design approach and emerging intelligent building operating systems such as digital platforms and control strategies. It will help Europe become an early leader in these promising technologies and will focus on building energy efficiency and the productivity and well-being of building occupants to renew the foundation for future competitiveness and growth in the coming decades.

Significant market acquisitions must be initiated to transform facade renovation technologies into mainstream technologies. This includes innovations in low-impact and climate-sensitive refurbished buildings through unexplored collaborations between advanced multidisciplinary design teams and innovative facade engineering. The European Energy Performance Building Directory (EPBD) highlights the importance of comfort, smart readiness, and high energy efficiency of buildings to be renovated [59]. With climate change and the increasing risk of overheating, the potential for the building renovation industry to renew the basis of its future competitiveness and growth and increase the use of facade refurbishment technologies and additional floor construction in the future is high. At this point, residents' attitudes toward new construction methods such as timber facade renovation and additional floor construction play a critical role in the spread of these practices and contribute to the transition to a forest-based bio-economy, the decarbonization of buildings, and a zero-energy building approach in Finland.

This study revealed the characteristics that residents value and guided those concerned about the direction in which residential areas should be developed. According to the results, residents were ready for large-scale use of wood in suburban development and renovation. The decision is ultimately made by the homeowners' positive attitude toward wood facade renovation, and the additional floor construction is an encouraging display of the enormous potential of the construction method. The additional floor construction contributes to overcoming the obstacles encountered in the difficult facade renovation process. Wood as a renewable material and carbon sink as a working material should be used for suburban regeneration as an ecologically responsive engineering solution to contribute to the decarbonization of buildings and a zero-energy building approach.

The results of this chapter were based on the empirical approach—residents' survey—but further research such as energy analysis, wood-based facade renovation, and additional floor solutions, developed prototypes testing will be done as part of other studies by following the methodology suggested in **Figure 6.**

Appendix: sample questions used in the questionnaire

- Answer the following questions related to Eskolantie wooden apartment buildings:

  a. How do you think Eskolantie's wooden apartment buildings are suitable for a residential area?

  * Very well * Well * Neutral * Poorly * Very poorly * I do not know.

  b. What do you think about the architectural features of the wooden apartment buildings in Eskolantie?

  * Very satisfied * Satisfied * Neutral * Dissatisfied * Very dissatisfied * I do not know.

  c. Do you accept new wooden additional construction in your residential area like Eskolantie wooden apartment buildings?

  * Agree * Partially agree * Neutral * Partially disagree * Disagree * I do not know.

  d. The wooden apartment buildings in Eskolantie have changed your opinion about wooden construction:

  * Positive * Partially positive * Neutral * Partially negative * Negative * I cannot say.

- The impact of Eskolantie wooden apartment buildings on the quality of the living environment in Pukinmäki (compared to the quality of the living environment before their construction) has been:

  * Positive * Partially positive * Neutral * Partially negative * Negative * I cannot say.

- If you wish, indicate that you think it is particularly successful in the wooden apartments in Eskolantie.

- If you wish, indicate that you think it is particularly unsuccessful in the wooden apartments in Eskolantie.

Author details

Markku Karjalainen, Hüseyin Emre Ilgın*, Lauri Metsäranta and Markku Norvasuo
Tampere University, Tampere, Finland

*Address all correspondence to: emre.ilgin@tuni.fi

References

[1] European Commission. A Renovation Wave for Europe—Greening Our Buildings, Creating Jobs, Improving Lives. Brussels: European Commission; 2020

[2] Majcen D, Itard L, Visscher H. Statistical model of the heating prediction gap in Dutch dwellings: Relative importance of building, household and behavioural characteristics. Energy and Buildings. 2015;**105**:43-59

[3] Filippidou F, Navarro JPJ. Achieving the Cost-Effective Energy Transformation of Europe's Buildings, the Joint Research Centre (JRC). Luxembourg: Publications Office of the European Union; 2019

[4] European Commission. New Rules for Greener and Smarter Buildings Will Increase Quality of Life for All Europeans. European Commission. 2019. Available from: https://ec.europa.eu/info/news/new-rules-greener-and-smarter-buildings-will-increase-quality-life-alleuropeans-2019-apr-15_en [Accessed: December 11, 2021]

[5] European Commission. Communication from The Commission to The European Parliament, The Council, The European Economic and Social Committee and the Committee of The Regions, European Commission, Stepping up Europe's 2030 climate ambition Investing in a climate-neutral future for the benefit of our people. Brussels: European Commission; 2020

[6] European Commission. Renovation Wave, Energy Efficient Buildings. 2021. Available from: https://ec.europa.eu/energy/topics/energy-efficiency/energy-efficient-buildings/renovation-wave_en [Accessed: December 11, 2021]

[7] Häkkinen T, Sustainable Refurbishment of Exterior Walls and Building Facades, Final Report, Part A—Methods and Recommendations, Julkaisija – Utgivare – Publisher VTT Technical Research Centre of Finland, Espoo, Finland. 2012. Available from: https://www.vttresearch.com/sites/default/files/pdf/technology/2012/T30.pdf [Accessed: December 11, 2021]

[8] BPIE (Buildings Performance Institute Europe). On the Way to a Climate-Neutral Europe – Contributions from the Building Sector to a Strengthened 2030 Climate Target. 2020. Available from: https://www.bpie.eu/publication/on-the-way-to-a-climate-neutral-europe-contributions-from-the-building-sector-to-a-strengthened-2030-target/ [Accessed: December 11, 2021]

[9] Shady A. Net Zero Energy Buildings (nZEB)—Concepts, Frameworks and Roadmap for Project Analysis and Implementation. Amsterdam, Netherlands: Elsevier; 2018

[10] European Commission. Nearly Zero-Energy Buildings. 2021. Available from: https://ec.europa.eu/energy/topics/energy-efficiency/energy-efficient-buildings/nearly-zero-energy-buildings_en [Accessed: December 11, 2021]

[11] BPIE (Buildings Performance Institute Europe). Positive Energy Buildings. 2021. Available from https://www.bpie.eu/event/positive-energy-buildings-wishful-thinking-or-built-reality/ [Accessed: December 11, 2021]

[12] Finnish Government. Carbon Neutral Finland that Protects Biodiversity. 2021. Available from: https://valtioneuvosto.fi/en/marin/government-programme/carbon-neutral-finland-that-protects-

biodiversity [Accessed: December 11, 2021]

[13] Simson R, Fadejev J, Kurnitski J, Kesti J, Lautso P. Assessment of retrofit measures for industrial halls: Energy efficiency and renovation budget estimation. Energy Procedia. 2016;**96**: 124-133

[14] Official Statistics of Finland (OSF). Building Stock 2020, Statistics Finland. 2021. Available from: https://www.stat.fi/til/rakke/2020/rakke_2020_2021-05-27_kat_002_en.html [Accessed: December 11, 2021]

[15] ARA. The Housing Finance and Development Centre of Finland, the Suburban Innova Block of Flats is Being Renovated into a Passive House. 2021. Available from: https://www.ara.fi/en-US/Housing_development/Development_projects/The_suburban_Innova_block_of_flats_is_be%2817681%29 [Accessed: December 11, 2021]

[16] Hirvonen J, Jokisalo J, Heljo J, Kosonen R. Towards the EU emissions targets of 2050: Optimal energy renovation measures of Finnish apartment buildings. International Journal of Sustainable Energy. 2019;**38**(7):649-672

[17] Kuittinen M, Häkkinen T. Reduced carbon footprints of buildings: New Finnish standards and assessments. Buildings and Cities. 2020;**1**(1):182-197

[18] Puuinfo. The Finnish Timber of Council. 2021. Available from: https://puuinfo.fi/?lang=enç [Accessed: December 11, 2021]

[19] Mills B, Schleich J. Residential energy-efficient technology adoption, energy conservation, knowledge, and attitudes: An analysis of European countries. Energy Policy. 2012;**49**:616-628

[20] Karjalainen M, Ilgın HE, Somelar D. Wooden additional floors in old apartment buildings: Perspectives of housing and real estate companies from Finland. Buildings. 2021;**11**:316

[21] Jurelionis A, Šeduikyte L. Multifamily building refurbishment process in Lithuania and other European countries. In: Proceedings of the 10th International Conference Modern Building Materials, Structures and Techniques. Vilnius, Lithuania: Vilnius Gediminas Technical University; 2010. pp. 106-111

[22] Botici AA, Ungureanu V, Ciutina A, Botici A, Dubina D. Sustainable retrofitting of large panel prefabricated concrete residential buildings. In: Proceedings of the Central Europe towards Sustainable Building 2013 Conference-Sustainable Building and Refurbishment for Next Generations; Prague, Czech Republic: CESB Prague SBE19 - SBE-Series; 2013

[23] Matic D, Calzada JR, Eric M, Babin M. Economically feasible energy refurbishment of prefabricated building in Belgrade. Energy and Buildings. 2015;**98**:74-81

[24] Bojić M, Miletić M, Malešević J, Djordjević S. Influence of additional storey construction to space heating of a residential building. Energy and Buildings. 2012;**54**:511-518

[25] Torres J, Garay-Martinez R, Oregi X, Torrens-Galdiz JI, Uriarte-Arrien A, Pracucci A, et al. Plug and play modular Façade construction system for renovation for residential buildings. Buildings. 2021;**11**(9):419

[26] Nicolae B, George-Vlad B. Life cycle analysis in refurbishment of the buildings as intervention practices in energy saving. Energy and Buildings. 2015;**86**:74-85

[27] Soutullo S, Giancola E, Sánchez MN, Ferrer JA, García D, Súarez MJ, et al. Methodology for quantifying the energy saving potentials combining building retrofitting, solar thermal energy and geothermal resources. Energies. 2020;**13**:5970

[28] Junghans L. Sequential equi-marginal optimization method for ranking strategies for thermal building renovation. Energy and Buildings. 2013;**65**:10-18

[29] Malacarne G, Monizza GP, Ratajczak J, Krause D, Benedetti C, Matt DT. Prefabricated timber façade for the energy refurbishment of the Italian building stock: The Ri.Fa.Re. project. Energy Procedia. 2016;**96**:788-799

[30] Attia S, Lioure R, Declaude Q. Future trends and main concepts of adaptive facade systems. Energy Science and Engineering. 2020;**8**:3255-3272

[31] Ruud S, Östman L, Orädd P. Energy savings for a wood based modular pre-fabricated façade refurbishment system compared to other measures. Energy Procedia. 2016;**96**:768-778

[32] Konstantinou T, Guerra-Santin O, Azcarate-Aguerre J, Klein T, Silvester S. A zero-energy refurbishment solution for residential apartment buildings by applying an integrated, prefabricated façade module. In: Proceedings PowerSkin Conference. Delft , Netherlands: TU Delft Open; 2017. pp. 231-240

[33] Patterson M, Vaglio J, Noble D. Incremental façade retrofits: Curtainwall technology as a strategy to step existing buildings toward zero net energy. Energy Procedia. 2014;**57**:3150-3159

[34] Bosman R, Rotmans J. Transition governance towards a bioeconomy: A comparison of Finland and The Netherlands. Sustainability. 2016;**8**(10)

[35] Karjalainen M. Status and possibilities of timber construction in Finland. In: Lilja K, editor. Wood-based Bioeconomy Solving Global Challenges. Helsinki: Ministry of Economic Affairs and Employment of Finland; 2017. pp. 35-39

[36] Karjalainen M. Multistory timber apartment buildings are becoming common in forested Finland—Results of resident surveys point the way to the future of multistory timber apartment buildings. In: Proceedings of 5th Annual International Conference on Architecture and Civil Engineering (ACE 2017); Singapore: Global Science and Technology Forum (GSTF); 2017. pp. 415-417

[37] Wood Building Programme. Finnish Ministry of Environment. Available from: https://www.ym.fi/enUS/Land_use_and_building/Programmes_and_strategies/Wood_Building_Program/Wood_Building_Programme(47800) [Accessed: December 11, 2021]

[38] Falk R. Wood as a Sustainable Building Material in Wood Handbook. Madison, WI: U.S. Department of Agriculture, Forest Service, Forest Products Laboratory; 2010

[39] Kremer PD, Symmons MA. Mass timber construction as an alternative to concrete and steel in the Australia building industry: A pestel evaluation of the potential. International Wood Products Journal. 2015;**6**(3):138-147

[40] Toppinen A, Autio M, Sauru M, Berghäll S. Sustainability-driven new business models in wood construction towards 2030. In: Filho WL, Pociovălișteanu DM, Borges de Brito PR, Borges de Lima I, editors. Towards a Sustainable Bioeconomy: Principles. Springer, Cham: Challenges and Perspectives; 2018. pp. 499-516

[41] Kuzman MK, Sandberg D, Haviarova E. Architects' perception of EWPs and modified wood in contemporary timber architecture. In: Proceedings of World Conference on Timber Engineering; Seoul, Republic of Korea; 2018

[42] Schuetze T. Wood constructions for sustainable building renovation. Advanced Materials Research. 2018;**1150**:67-72

[43] Kylkilahti E, Berghäll S, Autio M, Nurminen J, Toivonen R, Lähtinen K, et al. A consumer-driven bioeconomy in housing? Combining consumption style with students' perceptions of the use of wood in multi-story buildings. AMBIO. 2020;**49**:1943-1957

[44] Lähtinen K, Harju C, Toppinen A. Consumers' perceptions on the properties of wood affecting their willingness to live in and prejudices against houses made of timber. Wood Material Science and Engineering. 2019;**14**(5):325-331

[45] Larasatie P, Guerrero J, Conroy K, Hall T, Hansen E, Needham M. What does the public believe about tall wood buildings? An exploratory study in the US Pacific Northwest. Journal of Forestry. 2018;**116**:429-436

[46] Hu Q, Dewanckera B, Zhang T, Wongbumru T. Consumer attitudes towards timber frame houses in China. Procedia: Social and Behavioral Sciences. 2016;**216**:841-849

[47] Thomas D, Ding G, Crews K. Sustainable timber use in residential construction: Perception versus reality. In: Energy and Sustainability V, WIT Transactions on Ecology and The Environment. Vol. 186. Southampton, UK: WIT Press; 2014

[48] Høibø O, Hansen E, Nybakk E, Nygaard M. Preferences for urban building materials: Does building culture background matter? Forests. 2018;**9**(8):504

[49] Soikkeli A. Additional floors in old apartment blocks. Energy Procedia. 2016;**96**:815-823

[50] Simona D, Peter OTV. ProGETonE Public Deliverable D2.1: Report on the State of the Art of Deep Renovation to nZEB and Pre-Fab System in EU. Brussels, Belgium: European Commission; 2017

[51] BRESAER. (BREakthrough Solutions for Adaptable Envelopes in building Refurbishment). European Union's Horizon 2020 Research and Innovation Programme under Grant Agreement No 637186. 2020. Available from: http://www.bresaer.eu/ [Accessed: December 11, 2021]

[52] E2VENT. Energy Efficient Ventilated Facades for Optimal Adaptability and Heat Exchange, supported by the European Commission under the Energy Theme of the Horizon 2020 for Research and Technological Development. Grant Agreement Number 637261. 2018. Available from: http://www.e2vent.eu/ [Accessed: December 11, 2021]

[53] EENSULATE. Development of Innovative Lightweight and Highly Insulating Energy Efficient Components and Associated Enabling Materials for Cost-Effective Retrofitting and New Construction of Curtain Wall Facades, European Union's Horizon H2020 Research and Innovation Programme under Grant Agreement No. 723868. 2021. Available from: http://www.eensulate.eu/ [Accessed: December 11, 2021]

[54] P2ENDURE. Plug-and-Play Product and Process Innovation

for Energy-Efficient Building Deep Renovation, the European Union's Horizon 2020 Research and Innovation Programme under Grant Agreement No. 723391. 2019. Available from: https://www.p2endure-project.eu/en [Accessed: December 11, 2021]

[55] D'Agostino D, Zangheri P, Castellazzi L. Towards nearly zero energy buildings in Europe: A focus on retrofit in non-residential buildings. Energies. P2ENDURE. 2019;**10**:117

[56] Ferrante A, Prati D, Fotopoulou A. TripleA-Reno: Attractive, Acceptable and Affordable Deep Renovation by a Consumers Orientated and Performance Evidence Based Approach. WP4–Task 4.2 Analysis and Design of the Business Module. Maastricht, The Netherlands: Huygen Installatie Adviseurs; 2018

[57] D'Oca S, Ferrante A, Ferrer C, Pernetti R, Gralka A, Sebastian R, et al. Technical, financial, and social barriers and challenges in deep building renovation: Integration of lessons learned from the H2020 cluster projects. Buildings. 2018;**8**(12):174

[58] Metsäranta L. Residents' Opinions about Wooden: Facade Repair, Roof Extension and Urban Infill – Resident Survey to Pukinmäki District in Helsinki [MSc thesis]. Tampere, Finland: School of Architecture, Tampere University; 2020

[59] Hogeling J, Derjanecz A. The 2nd recast of the energy performance of buildings directive (EPBD). EU Policy News. REHVA Journal. 2018;**1**:71-72

Chapter 5

An Integrated Design Process in Practice: A Nearly Zero Energy Building at the University of Brasília - Brazil

Cláudia Naves David Amorim,
Joara Cronemberger Ribeiro Silva, Caio Frederico e Silva,
Thiago Montenegro Góes, Ayana Dantas de Medeiros,
João Manoel Dias Pimenta, Marco Antonio Egito,
Adolfo Bauchspiess, Loana Nunes Velasco
and José Manoel Morales Sánchez

Abstract

This study aims to present the design experience of LabZERO|UnB, an NZEB building awarded in a public call, that will be built on the University of Brasília campus. The method consisted of defining the design team and the Integrated Design Process (IDP), establishing assumptions and design guidelines, schematic design, initial computer simulations, design development, new simulations, and final calculations for the synthesis of energy performance. As a result, IDP proved to be efficient and underlined the possibility of translating research experiences into practice. The barriers and potentialities related to the coordination of a multidisciplinary team stand out, likewise the organization, planning, and achievement of goals. In the design concept of the 200m^2 building, the basic assumption was the adequacy of the architecture to favor the use of passive resources, respecting the local climate, classified as high-altitude tropical climate. Moreover, bioclimatic strategies were used, such as the North/South orientation of main façades, narrow floor plan, limited window-wall ratio, and adequate construction materials, to optimize energy consumption. As a result, the distributed generation of electricity was estimated at 58.29 kWh/m^2. a year and the final electricity demand was 34.29 kWh/m^2. year. Hence, this process indicates the real possibility of reaching the zero energy balance.

Keywords: nearly zero energy building, integrated design process, computational simulation, PV energy

1. Introduction

Buildings are a central part of the transition to a low-carbon society, with less environmental impact and energy efficiency, as they are responsible for consuming 32% of all energy generated in the world. This is equivalent to 19% of greenhouse gas emissions [1], in addition to consuming 50% of all raw material extracted by human action [2]. Initiatives such as the Sustainable Development Goals (SDGs) of the United Nations (UN) [3], the New Urban Agenda [4], and the Paris Agreement [5] point to the need for the reduction of energy consumption in buildings, generation of clean energy and more sustainable cities and communities to mitigate climate change and environmental crisis.

Buildings with zero energy balance, also known as the term Zero-Energy Buildings (ZEB), have an equivalent demand and generation of renewable energy within a year [6]. However, the equivalence between consumption and generation is not enough, because it is essential to achieve a demand reduction. For this, conservation and energy efficiency strategies are needed from the preliminary design [7], since they also provide thermal and lighting comfort, in addition to minimizing the environmental impact of the building in its operating phase. This includes integration with passive strategies, particularly in terms of natural lighting and ventilation, and high-performance enclosures. According to [8], new buildings have the potential to reduce energy demand by 50% if compared to the traditional ones, only by adopting commercially available energy conservation and efficiency strategies.

Thereby, the idea of NZEB emerged in the 1990s and afterward became part of energy policies in several countries. In Europe, the EU Directive on Energy Performance of Buildings [9] set goals to turn all buildings nearly zero-energy by 2020. The US Department of Energy's Building Technologies Program established similar objectives: achieving zero energy homes by 2020 and zero-energy commercial buildings by 2025 [10]. In addition, this building category is aligned with the 7 and the 11 UN Goals of Sustainable Development. According to O´Brien et al. [7], the NZEBs are characterized by a rigorous design and operation of the building as an integrated energy system, with a good indoor environment suited to its role. Some key points are mentioned, such as: an integrated approach to energy efficiency, passive and active design and building operation; optimization of solar collection, requiring building design and roofs used for conversion to electrical energy, useful heat, and natural lighting. **Table 1** shows the difference in project design and operation between conventional buildings and NZEB buildings.

In Brazil, there is still no regulation regarding NZEBs, but specific actions have been taken to leverage the improvement of energy efficiency in buildings, through regulations and standards [11], building performance [12], and distributed energy generation [13]. However, concrete actions for the construction and monitoring of NZEB buildings are recommended to enable the dissemination of the concept. In this context, the National Program for Energy Efficiency in Buildings (Procel Edifica) carried out a Public Call in 2019 to support the construction of up to 4 (four) NZEB's in strategic locations throughout the country [14]. The objectives of the call included: to foster knowledge, research, and development of NZEB project designs; to create a demonstration effect of NZEB buildings, enabling large-scale adoption, and, finally, to verify the technical and financial feasibility of the construction and operation of NZEB buildings. The Public Call was launched on December 2nd, 2019, and the deadline was set to February 20th, 2020. The call requested the submission of the Basic Project Design of the NZEB new construction or to undergo retrofit, bringing

DOI: http://dx.doi.org/10.5772/intechopen.102443

Design and operation of building systems	Conventional buildings	NZEB buildings
Building envelope's materials	Passive, not designed as an energy system	Optimized in passive design integrated with active solar systems
Heating and ventilation air conditioning (HVAC)	Large systems, oversized	Optimized small HVAC systems, integrated with solar systems, combining heating and power, seasonal storage, and district energy
Solar systems / renewable energy technologies (RET)	No systematic integration – an afterthought	Fully integrated: natural lighting, solar thermal, photovoltaic, hybrid, geothermal, biofuels integrated with smart microgrids
Building automation systems	Building automation system not effectively used	Building Controls for optimizing performance
Design and operation	Design and operation are considered separately	Fully integrated and optimized design and operation, considering environmental comfort

Table 1.
Design and operation of NZEB buildings versus conventional buildings.

together "the elements that define the building, aiming at the accuracy of its basic characteristics and its desired performance in the work, with the estimated cost and execution time" [14].

The University of Brasília (UnB) has been investing in strengthening sustainability actions on its campuses; according to Taucher and Brandli [15] (2006) "the socio-environmental dimension, in this context, stands as a principle for institutional development". Thus, the construction of a zero-energy balance building and possibly replicable typology proves to be an important step towards the dissemination and consolidation of sustainable practices at the University, with positive consequences and impacts even for the city. Therefore, to advance on sustainability purposes, UnB's multidisciplinary team developed a project design for a laboratory and coworking space, called LabZERO|UnB, which was one of the 4 buildings included in the so-called Procel Edifica Public Call (3rd place overall).

This study presents in detail the design process experience of this NZEB building - initially, all design phases, results, barriers, and potential are addressed. Afterward, the final design and the analysis of environmental and energy performance are presented, and the challenges to the implementation of this type of practice, and the relevance of initiatives to promote the dissemination of zero-energy balance buildings, are discussed.

2. Research and design for an NZEB

2.1 Integrated design process of an NZEB building

Note that the characteristic of this type of building involves a project that integrates passive and active systems, in addition to the specification of optimized ventilation and air conditioning systems, connecting natural light and power generation. On the other hand, design practice must shift from a traditional linear process to a collaborative approach between architects, structural engineers, mechanics, electricians, and other professionals. By definition, the Integrated Design Process (IDP) guides decision-making in various professional specialties, including the use of natural resources, energy consumption, and the achievement of environmental quality [7, 10]. Kwok and Grondzik [16] define the IDP as one that synergistically involves several disciplines, to create more efficient and responsible buildings with a lower life cycle cost. Keeler and Burke [10] conceptualize it as a synonym for sustainable design. The authors emphasize that in the case of integrated design, it is important to understand the design variables as a unified whole, involving decisions about energy consumption, natural resources, and environmental quality.

The main features of the integrated project are:

Iterative, non-linear process: In contrast to the conventional (linear) design process, in which team members work in isolation, PPI promotes ever-increasing feedback loops among everyone;

Collaboration and innovation: All participants must share the same vision of the project from the beginning, in order to provide input and feedback to the rest of the team. Project contributors may be asked to work on tasks outside their usual objective. PPI encourages everyone to share the learning and improve the process as a whole;

Multidisciplinary team: Ideally, the PPI includes all stakeholders in a project, and they must be present from the early stages of the work, providing their expertise to the project process. There may be other consultants, depending on the specific needs of each project [7].

These authors mention strategies and aspects related to the design of the NZEB building, by pointing out the design issue and listing the iterative phases of design in **Conceptual Design**, **Project Development,** and **Technical Design**, as shown in **Figure 1**.

Another aspect mentioned by the same authors is related to technical and research matters, in which they highlight the computational model simulation that is going to be used. The importance of research inputs to be applied during the design process is also emphasized. In order words, the development of an NZEB project requires prior knowledge and research, especially in cases of restricted deadlines. Monteiro et al. [17] state that in this type of project, computer simulation has become a mandatory step in the process, adding complexity, but favoring the improvement of the project.

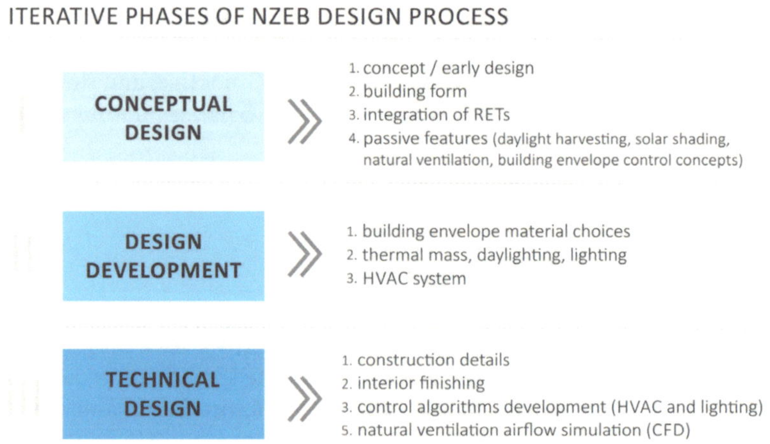

Figure 1.
Iterative phases of the NZEB design process. Source: Adapted from [7].

Mendes and Amorim [18] report an experience of applying the concepts of Integrated Project in a graduate discipline, during which the method proposed by O'Brien et al. [7] was used with two crucial factors: well-defined project objectives shared by the entire team and the presence of a facilitator (coordinator), who sets the tone for collaboration and effective communications during the design process. There was also the creation of teams of specialists in the various themes to be addressed in the project, along with the establishment of periodic meetings with the entire group to share results and align actions. A team specialized in computer simulations acted transversally, receiving and providing inputs to the others. The experience proved to be efficient, noting that the design process reached appropriate fluidity and the project proposed in the discipline achieved appropriate technical results, with an energy consumption lower than its production, reaching the goal of becoming an energy balance building null [18]. This was defined as the basis of the method to be used in the LabZERO|UnB integrated design experience.

NZEB design, monitoring, and benchmarking experiences reported by Garde and Donn [19] present 30 residential and non-residential case studies, grouped into cold, moderate, and hot climates. Three of these buildings can be compared to the conditions of LabZERO|UnB due to the similarities in use and climatic conditions. In these cases, energy demands ranging from 16 to 66 kWh/m^2.year are identified, with energy production ranging from 44 to 115 kWh/m^2.year. In one case, energy production is 7 times greater than the demand. **Table 2** presents energy demand and production data.

2.2 Team definition, initial guidelines, and preliminary design

The definition of the team is an important part of the project conception, as their profile must be able to provide the full development of the products, within the stipulated time limit. It also established the involvement of administrative bodies linked to the project and construction management of the University, as it is a proposal for the construction of a building on the campus, involving bureaucratic issues and administrative actions. Furthermore, the expertise of technicians linked to the university's construction sector is essential for the development of the project in accordance with internal rules. More than that, the technicians carry out theoretical work, resulting from research in the area, and act at the same time in training regarding the bioclimatic project, energy efficiency, etc. This partnership between research and project/action is seen as crucial to leverage more effective actions towards greater efficiency in construction on the University campuses as a whole. In conclusion, there is a need for a mixed team that combines a variety of researchers and professionals from different

Building and location	Typology	Energy demand Kwh/m^2.year	Energy production KWH/m^2.year
ENERPOS- La Réunion (21°S, 55°L)	Offices and classrooms	16	115
Illedu Centre - La Réunion (21°S, 55°L)	Offices	66	92
ZEB@BCA Singapure (1°20 N, 103°L)	Offices and classrooms	40	44

Table 2.
Energy demand and production in 3 NZEBs. Source: [19].

specialties and modes of activity, able to apply the concepts of previous research and work developed by teams of professors and researchers and implement them in the project proposal in an agile way.

Once defining the team, meetings there will be meetings to take preliminary decisions regarding the nature and size of the project, considering budget and deadline limitations. Other decisions taken preliminarily are related to the type of the building (residential, commercial), function, and location on the University campus. According to the methodology proposed by O'Brien et al. [7], the team facilitator should have the task of delimiting attributions for each of the participants and defining the delivery deadlines, depending on the necessary feedback from each phase of the project. The technical drawings required by the contest announcement were: topographic survey; location and situation plan; architectural project; hydraulic installations design; electrical installations; air conditioning; lighting; and distributed generation project from renewable sources. Besides the Basic Project, there were other mandatory items to be delivered, such as Requirements of Use, Descriptive Memorandum, Budget, Schedule, Energy consumption, and distributed generation evaluation report and Preliminary Visitation Plan. It is noteworthy that the building must be open to visitation and monitored within 24months of its construction, to allow the measurement of its real performance.

The preliminary design of the building was done with a defined area due to budget constraints. Initial decisions and common goals must be developed with the participation of all.

According to the premises established in the methodology, the participants chosen were members of the research groups and laboratories at the University of Brasilia and those working closely with the NZEB theme and disciplines related, such as the postgraduate course Integrated Environmental Project, created in 2017 and taught in the Postgraduate Program in Architecture and Urbanism at the University of Brasília. This core team is coordinated by professors of the Architecture and Urban Planning -(Laboratory of Environmental Control and Energy Efficiency - LACAM), alongside with professors of Mechanical Engineering (Air-Conditioning Laboratory - LaAr) and Electrical (LARA - Automation and Laboratory) Robotics), partners since 2014 in the development of disciplines, undergraduate and graduate final works on the subject [20–22]. Professors of Geology and Environmental Science were also involved to develop themes related to the project's sustainability (water, waste, etc.).

The team was defined with 24 members, as follows: 2 architects specialized in energy efficiency, process coordinators; 2 specialists in a computer simulation, who transit between all other teams; 1 architect specialized in energy efficiency; 3 architects and 1 civil engineer without training in energy efficiency; 1 mechanical engineer specialized in energy efficiency (responsible for HVAC); 2 specialists in electrical engineers (1 responsible for photovoltaic energy generation, the other for controls and automation); 2 engineers specialized in budgeting; 2 engineers specialized in the use of water and waste; and 4 undergraduate students in Architecture. There was also the collaboration of a company residing in the University's Science and Technology Park, a specialist in energy efficiency labeling in buildings, and a junior company active in the field of civil construction, composed of graduate students in Civil Engineering and experts in the preparation of budgets for construction.

It was initially considered to use an NZEB residential building project, the result of an existing master's dissertation [22], but impasses regarding the use and occupation of a residential establishment on a university campus, in particular related to security and monitoring, eliminated this proposal. The second hypothesis dealt with the use of a retrofit project, carried out previously [15], in an existing building on the campus. In this case, the limiting factor was the cost, since it is a large building, the budget would exceed the amount offered by the Public Call. The Birck project [20], previously mentioned, due to its large area would also present a high cost. It was therefore decided to carry out a new building project on the campus.

After the initial discussions, the project's objective was defined as follows: to build an open and collaborative laboratory, which would allow for some flexibility in the plant without specific programmatic needs.

The city of Brasilia, where LabZERO|UnB will be constructed, is located in the central area of Brazil (latitude 15°46'South and longitude 47°55′ West) (**Figure 2**) and it has a climate that is classified as high-altitude tropical climate or Tropical savanna climate (*Aw*, according to the Köppen climate classification), milder due to the elevation (1.100 m). This climate is characterized by a rainy season, from October to April, and a dry season, from May to September. The average temperature is 21.0 C.

Initial design guidelines included local climate recommendations in bioclimatic zone 4 as per ABNT 15220 [23], which indicates shading, controlled natural ventilation, light and insulated roof, limited window-wall ratio, and light colors. Additionally, a floor plan with reduced depth was defined to favor natural lighting and it was installed with the largest façades facing North and South, to reduce the incidence of sunlight and optimize the protection of the façades. The roof houses the photovoltaic panels, as well as the North façade, which receives photovoltaic brises

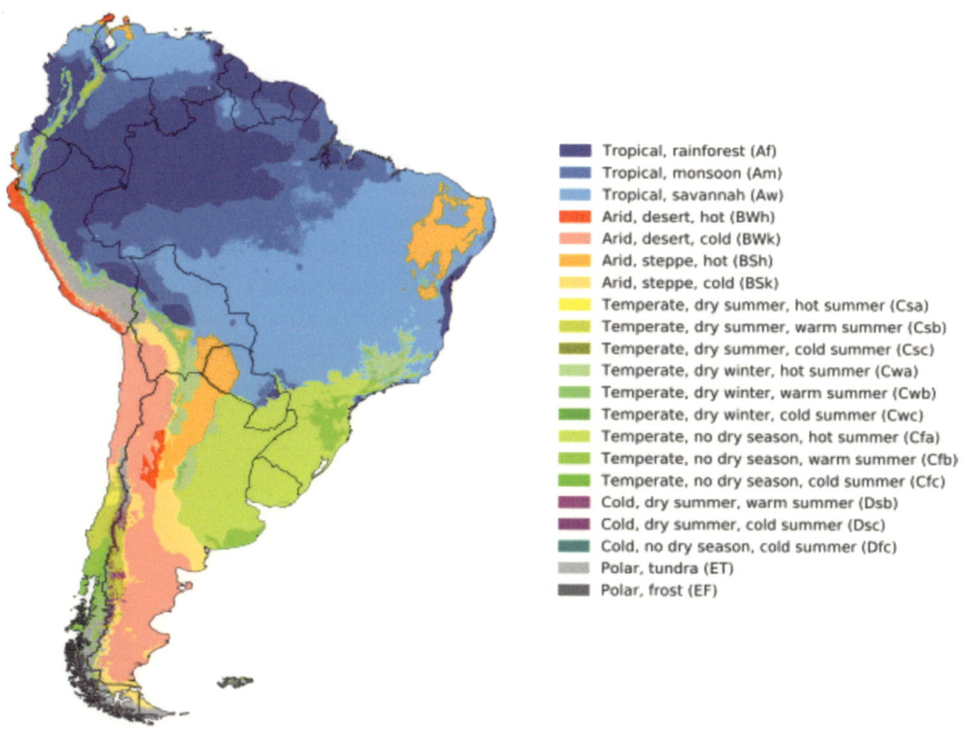

Figure 2.
Koppen-Geiger classification map for South America. Source: Beck, H.E., Zimmermann, N. E., McVicar, T. R., Vergopolan, N., berg, a.; wood, E. F.; present and future Köppen-Geiger climate classification maps at 1-km resolution nature scientific data. DOI:10.1038/sdata.2018.214., CC BY 4.0, https://commons.wikimedia.org/w/index.php?curid=74674070.

that also work as solar protection. The first design sketches (**Figure 3**) were developed based on these guidelines, but they gradually evolved as a result of discussions of the various aspects with the entire team. It is worth noting that the design process sought to harmonize esthetics with the local context of the university campus.

2.3 Simulations and preliminary draft

Computer simulations are carried out after the definition of the preliminary design to validate the first decisions regarding the implementation and orientation, the form of building, glazed area, solar protection systems, solar exposure for solar and photovoltaic panels. Some design variations and sensitive variables that feedback into the design process are tested in an integrated design action, in which team members participate. This process takes several weeks until an ideal energy solution is obtained.

To assess the building's energy performance, the Energyplus software was used through the DesignBuilder graphical interface for a period of a typical year. The results are presented by the energy consumption in kWh/m^2.year. The same software is used to perform the passive potential performance of the building's coworking area. In this case, the results are checked by the percentage of hours occupied in comfort using the adaptive comfort model of ASHRAE-55 for both 80% acceptability and 90%. As for the evaluation of the luminous performance of the coworking area, the Radiance program is used, through the Rhinoceros 3D program and its visual programming language Grasshopper and the add-on HoneyBee. The Daylight Autonomy (DA) is evaluated at 300 lux, and the Useful Daylight Illuminance (UDI) above 2000 lux.

The Basic Project, which is the level that the NZEB building proposal should be delivered for the PROCEL EDIFICA 2019 call notice [14], was defined after some alternatives were tested by simulation, in particular regarding sun protection, types of glass (light transmission and solar factor) and building materials (roofing and walls). In this phase, automation, and control strategies (HVAC and lighting), location of photovoltaic panels, such as Renewable Energy Technology (RET), lighting design, and other sustainability strategies, such as rational use of water and waste treatment, were also defined by teams of engineers and experts. The team participated in an integrated way. The group responsible for the simulations brought about results, which were evaluated under different aspects (energy, esthetic, functional, cost) before taking the final decision on the project.

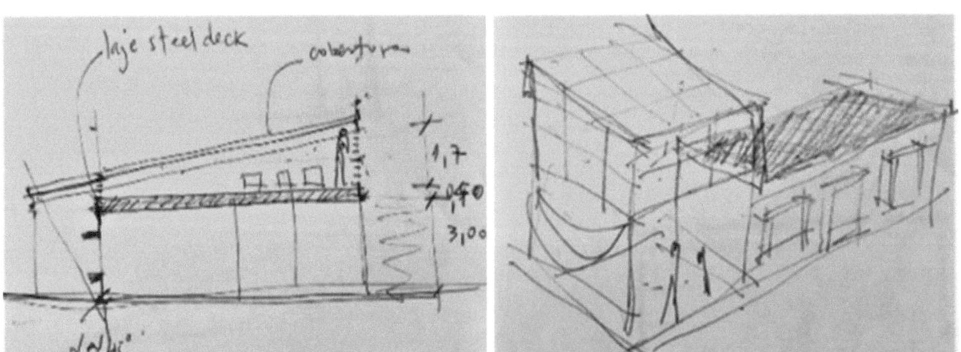

Figure 3.
Sketches with the first preliminary design risk, later revised (plan, volume, and section). Source: Authors.

Due to the first thermo-energetic simulations and daylighting, the preliminary design of the building was established.

After another round of simulations, the Basic Project was defined, bringing details of the preliminary project, such as envelope materials with thermal transmittance and absorptance suitable for the bioclimatic context (external walls, fiber cement panels, rock wool insulation, and plasterboard, U = 0.89 $W/m^2.K$; steel deck slab coverage, metallic tile, and insulation, U = 0.57 $W/m^2.K$); artificial lighting system with efficient lamps, luminaires, and task lighting; and automation for HVAC and artificial lighting.

2.4 Simulations and final calculations

After the definitions of the Basic Project, the feedback from the initial simulations, and the tests of several hypotheses, the final simulations of energy consumption involved the same software mentioned above. In addition to these, the RELUX software was used for simulations of the lighting project, the SAM software of the National Renewable Energy Laboratory (NREL) for dimensioning and calculation of two independent photovoltaic systems: on-grid and off-grid. Finally, energy efficiency labeling calculations, primary energy consumption, and budgets for final solutions, required by the notice, were performed. Regarding the budgets, a junior civil engineering company was counted on, which made the quotations of 21 items, plus the percentage of BDI, according to the model of the Public Call [14].

The simulations and final calculations prove that the building achieves an average annual consumption of electrical energy of 34.29 kWh/m2. year (7099.18 kWh/year), which corresponds to a primary energy consumption value of 54.88 kWh/m^2.year (11,358.68 kWh/year), that is significantly lower compared to the average consumption of electricity in office buildings in Brasília, which is around 130 kWh/m2.year [24]. As for the distributed generation of electricity in the photovoltaic system installed on the roof and side area, the value obtained is 58.29 kWh/m2.year. The results are consistent with international experiences in similar climates, presented above (**Table 2**). With these data, the achievement of the goal of building NZEB, or energy balance close to zero or nil, is proven.

The building's reduced energy consumption is achieved through architectural and technological strategies (passive and active). In addition to aspects of energy efficiency and comfort, the building proposes strategies for the rational use of water and waste management. Sustainability aspects are also highlighted, such as the steel structure and the sealings in prefabricated panels, allowing for quick and clean construction, with less waste generation and possible replicability of the typology.

The building obtained a level A energy efficiency label (the higher efficiency level, according to Brazilian National standards) as expected due to the inclusion of bioclimatic and energy efficiency strategies since its conception. In isolation, the envelope obtained EqNum = 5, the lighting obtained EqNum DPI = 5, and the air conditioning EqNumVent = 5, related to the Coefficient of Performance (COP) of the machines, with partial level A labels being obtained individually. As a bonus, it was counted the rational use of water (40% savings) and the energy savings from the network (more than 30%). The general prerequisite of dividing electrical circuits was also fulfilled. Therefore, the overall energy efficiency label obtained for the building is level A.

2.5 Design process: synthesis, limitations and potentialities

As indicated by the literature [3], the architecture started along with the conception of Renewable Energy Technologies (RET), which, in the present case, consisted of photovoltaic energy. Feedback cycles took place periodically between the thematic teams, together with the facilitators. The design of the building for the use of natural lighting also took place from the preliminary design. Soon after, the HVAC project, starting with passive strategies (ventilation, evaporative cooling, solar chimney) was initiated. The active HVAC strategies were designed right after the first thermo-energetic simulations, due to the hours of discomfort not passively resolved, giving rise to the preliminary project. At this point, an initial calculation of the building's energy balance was carried out (with data from the first thermo-energetic simulations and the photovoltaic panels still only on the roof). Then the sunshades and openings were readjusted to correct some identified problems. As a result of natural lighting simulations, with the building being better defined, the lighting, electrical, controls, and automation projects were carried out. In this preliminary project phase, strategies were also conceived for the rational use of water (hydro-sanitary project) and waste treatment, which are complementary aspects of the project's sustainability. The final thermo-energetic simulations, labeling, primary energy calculations, and final energy balance of the building were carried out after the definition of the envelope materials, the internal finishes, and the basic project, **Figure 4** presents the design process, products, and flows, relating them to the iterative phases mentioned in **Figure 1**.

The process took place relatively smoothly, due to the aforementioned previous tests, involving part of the team. However, some important points that emerged from the experience with time and budget limitations should be mentioned: 1. The role of the facilitators is essential to coordinate the various decisions to be taken that

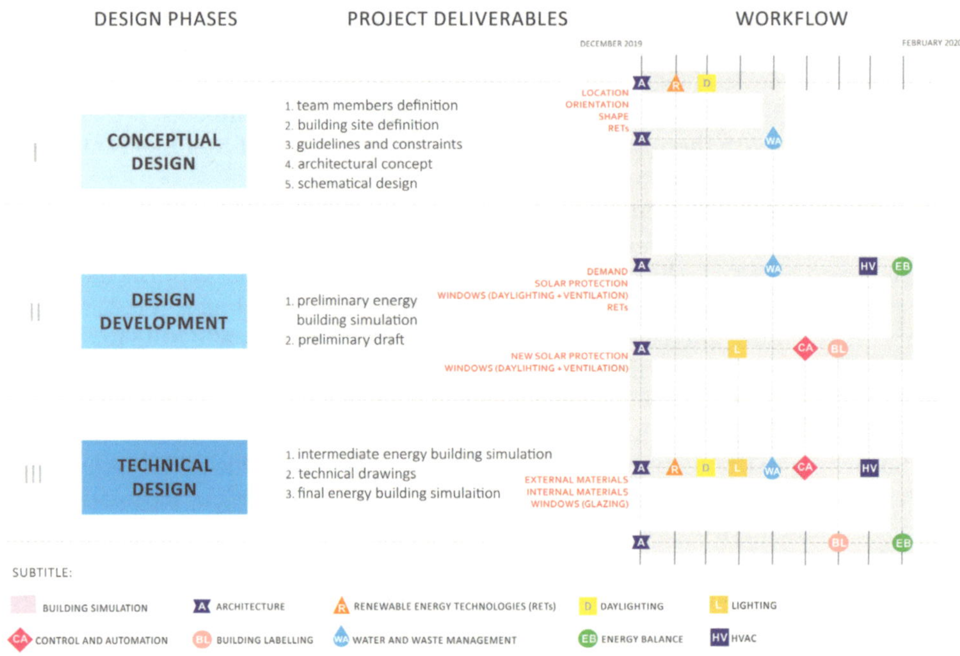

Figure 4.
Design process, with products and design flow. Source: authors.

require inputs and results from different thematic teams; good facilitators are crucial for meeting deadlines and are potential drivers of positive results; 2. Efficient communication with the various thematic teams is included in the role of the facilitators, to delimit the level of detail of the solutions proposed by each one, in each phase. In the early design phases, the level of detail should be lower, to avoid wasting time and rework; in the final stages, the level of detail is higher. There seems to be a tendency among specialists to get the phases in detail from the beginning., to be controlled by the facilitators, as it represents a barrier to the fluid development of the process; 3. The simulation team also has a fundamental role and interacts with other teams, as they need to "translate" the architectural proposals into simulation results, which feedback the new architectural proposals. For this, communication must be effective, and the language adapted to reach all professional profiles, which is not simple and can become a barrier in the process; 4. Periodic meetings, sharing information, and decisions are important for team involvement and motivation. However, in some moments, quick decisions must be taken and for this, again, the role of the facilitators is fundamental.

3. The LabZero UnB project: Design, performance analysis and energy balance

This section presents the final design of LabZERO|UnB building, as a result of the previously described design process. The performance analysis, computational simulation process, and final energy balance are also presented.

3.1 The final design of LabZERO|UnB

The LabZERO|UnB construction is predicted to be done at the Science and Technology Park, at the Darcy Ribeiro campus of the University of Brasília (UnB), which aims at socio-economic development and strengthening research, development, and innovation (RD&I) structures. The privileged location on the campus provides the building with excellent visibility and easy access for the visitors (**Figure 5**).

Once built, the LabZERO|UnB building will be used for office activities in a coworking regime, to house research groups of UnB's Architecture and Engineering Faculties dedicated to the study of zero energy balance and sustainability in buildings, (**Figure 6**).

In terms of architectural design, the basic assumption was the adequacy of the architecture to favor the use of passive resources, respecting the local climate recommended strategies for bioclimatic zone 4 (Bioclimatic Zone 4, [23]), which includes shading, controlled natural ventilation, roof insulation, among others, as mentioned before in 2.2.

It was also a premise that architectural style was in accordance with the construction standards of the University of Brasília, highlighting, in the volumetry, some of the innovative systems used in the building.

Considering the educational and representative character of LabZERO|UnB, both internally and externally, the architecture uses innovative systems as elements of a visual framework, to highlight the applied design decisions, such as the steel structure, apparent electrical installations, and visual integration between the technical area and the work environment. As for the building's morphology (**Figure 7**), the elongated and shallow shape, with larger façades towards the North–South orientation, allows the use

Figure 5.
Location of UnB campus Darcy Ribeiro in Brasilia.

Figure 6.
Building plot on the Darcy Ribeiro campus (left) and implementation (right). Source: [25].

of natural light and optimized and effective sun protection [25]. The glazed area on the façades is limited to 35% and duly protected from excessive sun radiation using louvers. On the North façade, they are indeed a BIPV (building-integrated solar photovoltaics) solar louvers, whilst a solar chimney system is present on the West façade to intensify natural ventilation, combined with forced ventilation when necessary (**Figure 7**).

The floor plan has 207 m^2, arranged as an office area (with an area reserved for meetings), a pantry, a bathroom, a dressing room, a technical area, and a bicycle rack, in addition to an outdoor balcony. **Figure 8** shows the layout of the building plan.

The constructive systems elected to be used in this project are envisaged to strengthen sustainability and technological innovation. In addition, institutional criteria had to be met regarding the possibility of reproducibility, relocation, and integration with industrialized dry-construction systems, which reduce losses and

Figure 7.
3D perspective view of LabZERO|UnB building from northeast [25].

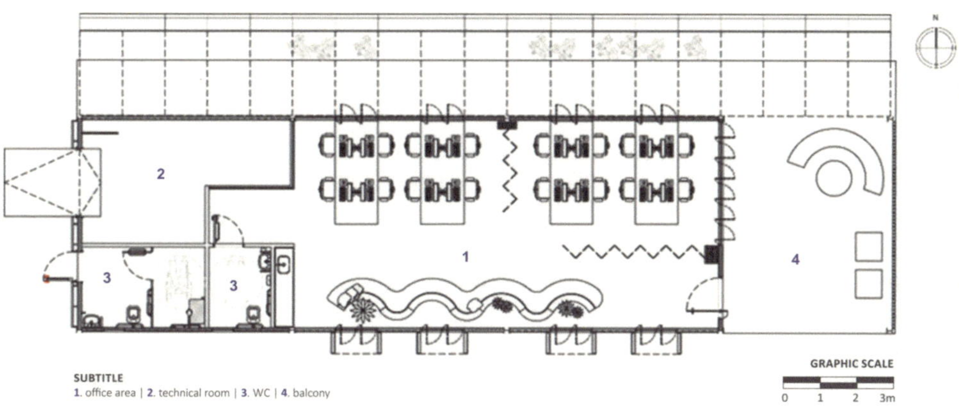

Figure 8.
Floor plan of LabZERO|UnB building. Source: [25].

waste in construction, ensuring faster execution. The building envelope systems are composed of a composite steel deck slab plus 12 cm of concrete employed on the floor and the roofs, whereas external walls are constituted by external fiber cement panel and drywall internally, filled with 4 cm of rock wool for insulation. Internally, all partitions are composed of two drywall panels with an air cavity, except for the partition between the office area and the technical area, which employs a clear 6 mm glass.

As complementary processes, in addition to natural lighting and ventilation, it was included an induced (or forced) ventilation system using a solar chimney. When comfort conditions with natural ventilation and induction were not sufficient, a set of high-efficiency exhaust fans with speed control is activated, maintaining the necessary airflow for the occupied space. Additionally, to the several passive systems and techniques envisaged to maintain thermal and lightning comfort conditions, the building's energy efficiency is guaranteed by highly efficient artificial lighting and HVAC appliances. The project will have a rational use of drinking water, besides the use of alternative water sources, distributed generation with grid-connected photovoltaic generators, waste management, accessibility, and new technologies (**Figure 9**).

From the early stages, the building was conceived to achieve high performance and renewable energy generation instead of contemplating only conservation, efficiency, and energy generation measures in the final stages of the project. This is especially relevant because it is in the initial design stage that there is the opportunity to reduce the project costs and avoid future rework [26]. However, in order for this to happen, the project methodology contemplated interaction and collaboration between the various agents and disciplines that interfere in the project development, which in fact occurred in the experience of LabZero at the University of Brasília (LabZERO|UnB).

3.2 Performance analysis guidelines

Several aspects of the project were evaluated using computer simulation tools, not only to estimate electricity consumption, generation demands, and comfort conditions, essential for the development of a zero-energy balance building project but also to support the decision-makers in design. The computer simulation tools also helped to envision the building's tagging process. In this section, the main guidelines and assumptions for environmental and energy performance analysis of the LabZERO|UnB project are presented.

3.2.1 Daylighting

For daylighting analysis, the Radiance program was used through the Daysim/Honeybee graphic interface, and Grasshopper/Rhinocerous3D plugin (**Figure 10**). To evaluate the performance of daylighting, 2 metrics were used: DA (Daylight Autonomy – or Natural Lighting Autonomy) considering 300 lux, and UDI (Useful Daylight Illuminance) considering a maximum of 2000 lux. In both cases, the measurement plane considers the height of the work plane at 80 cm in relation to the floor and the mesh of stitches distributed every 50 cm. In terms of the availability of natural light during the period of occupation of the building, the interval from 8 am to 6 pm was considered valid, during all 12 months of the year.

3.2.2 Lighting

Artificial lighting was designed considering daylighting availability. The computer simulations used to verify the condition of artificial lighting in the building were

Figure 9.
On the left, the perspective shows the BIPV solar louvers on the north façade and a solar chimney on the west façade; on the right, the perspective of the south façade source: [25].

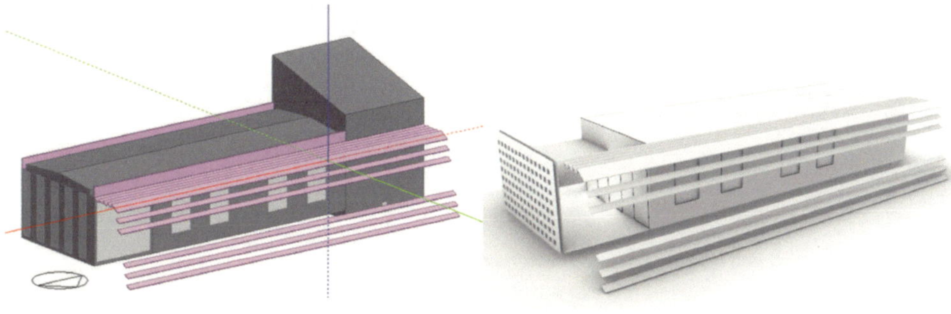

Figure 10.
On the left, modeling in DesignBuilder and on the right modeling in rhinoceros 3D [25].

carried out by modeling and calculating the data through the/Relux software, version 2019.3. The objective was to optimize the energy efficiency of the system, ensure adequate lighting rates, according to [27], and serve as a basis for an energy assessment. The input data were: the building geometry (height, width, depth, and useful ceiling height); the artificial lighting equipment in each environment; and the height of the work plane (70 cm).

3.2.3 Thermo-energetic performance

To analyze the building energy performance and to verify the electricity demand, EnergyPlus 8.9 was used, through the graphic interface DesignBuilder 6.0 (**Figure 10**). The model's geometry followed the architectural design, and the climate file was a Swera type for the city of Brasília-DF.

The loads and schedules utilized are based on the ASHRAE Handbook of Fundamentals [28] mostly for generic office area, which is the predominant occupation. The office breakroom in the outside area and bathrooms follow also the same indications [25], however, they are adapted to the Brazilian reality, so no plug load is considered in these areas. Additionally, since the technical area does not have heavy machinery, instead of the 52 W/m^2 considered to this kind of area in the ASHRAE Handbook of Fundamentals (2017), it is employed the same value of generic office area, of 11 W/m^2, which allows a general load closer to the generic offices found in Brasilia by Costa et al. (2018). The attic is considered unoccupied with no internal loads. Furthermore, the artificial lighting energy values are obtained from the lighting design, with an overall 5 W/m^2 for all environments, meanwhile, in the office area, there is an additional 1 W/m^2 for task lighting. **Table 3** summarize these data:

Additionally, the building operation varies from 8 h to 22 h on weekdays and all schedules are derived from this operation period. The simulation is carried out for the whole year and the data analyzed is Energy Use Intensity (EUI) in kWh.year/m^2, considering only the occupied area (not including the attic).

The reflectance of materials is based on the general guidelines of [12], which defines absorbance values for light colors as 0.4 and for dark colors as 0.7. The floor and the ceiling were modeled as dark, while other surfaces were defined as light.

The building envelope thermal properties follow the standards of [23], with [29] reference for modeling in EnergyPlus. The external vertical sealing composition comes from [22], external walls of fiber cement and rock wool (0.89 W/m^{2}K), in addition to a covering composed of metallic tile with insulation (0.80 W/m^{2}K), ventilated cavity

Item	Office	Breakroom	Water closet	Technical area	Attic
Occupation (person/m^2)	0,1100	0,2889	0,1124	0,1110	—
Equipment (W/m^2)	11,77	—	—	11,00	—
Artificial lighting (W/m^2)	5 + 1	5	5	5	—

Table 3.
Occupation, equipment, and lighting power per area type.

(10 ren/h), and steel deck slab (3.16 W/m^{2}K). The thermal properties of all layers of the opaque envelope are presented in **Table 4**.

The glass employed on the windows is a clear laminate 13 mm glass (6 mm+1 mm PVB+6 mm) (**Table 5**). All windows have external shading elements, as recommended for this climate.

As for electrical equipment, the installed power follows the RTQ-C as a reference [11], except for lighting that respects the project presented in the analysis of the artificial lighting system. The usage routine is from 8 am to 10 pm 5 days a week. With the exception of the coworking area, the other areas have natural ventilation. Bathrooms, technical area, and balcony have the ventilation network model (airflow network). According to the project, the frames opening rate is 88%.

For the attic zone, a constant rate of 10 renewals per hour is used. The office working area will be equipped with a highly efficient direct expansion HVAC system for cooling purposes. No heating will be employed since it is most frequently necessary late at night when there is no occupation in the building. It is employed ideal air loads for the mechanical systems with a Coefficient of Performance (CoP) of 5, which is a theoretical constant value for the equipment employed. There is also a cooling

Systems	Layers	Width (cm)	Conductivity (W/m.K)	Specific Heat Capacity(J/kg.K)	Density (kg/m^3)	U-Value (W/m^{2}K)
Steel Deck Slab	Steel	0,6	55,000	460	7800	3,16
	Concrete	12,0	1130	1000	2000	
Double Metal Roofing with Insulation	Steel	0,6	55,000	460	7800	0,47
	Rock Wool	9,0	0,045	800	100	
	Steel	0,6	55,000	460	7800	
External Wall	Fiber Cement Siding	1,0	0,950	840	550	0,89
	Rock Wool	4,0	0,045	800	100	
	Drywall	2,0	0,350	870	900	
Internal Partition	Drywall	2,0	0,350	870	900	1,80
	Air Cavity	11,0	Fixed R-Value of 0,18 m^2.K/W			
	Drywall	2,0	0,350	870	900	

Table 4.
Opaque envelope thermal characterization.

Characteristics	Clear glass 6 mm
SHGC (W/W)	0.74
Light transmission (W/W)	0.86
U-value (W/m^2K)	5.29

Table 5.
Glass thermal properties.

setpoint of 24 °C operative temperature with no setbacks. Finally, a water condensing unit is used in combination with an evaporator fan, which blows cold air from a plenum under the floor of the working area.

In addition, there is artificial lighting control in this zone, with setpoints of 150 lux for the balcony area and 300 lux for the coworking area.

3.2.4 Potential for photovoltaic energy generation from on-grid and off-grid systems

To analyze the potential of photovoltaic energy generation, the SAM software from the National Renewable Energy Laboratory (NREL) was used. Two different photovoltaic systems were designed. The first one was a photovoltaic field of a kind that is connected to the public distribution network (on-grid), integrated to the coverage of the technical area of the building, facing North, with an inclination of 15°. The other photovoltaic system was conceived as an integrated field to the design of the brise-soleil that shade the North façade – using a battery bank for storage (off-grid), and will not be directly connected to the public grid. This unusual design is intended to address future research regarding demand energy management.

A system with 12 TRINASOLAR TSM-DE15MII-400 W TALLMAX modules of 400 Wp of monocrystalline silicon was considered for the on-grid system and YINGLI YL100P-17B 2/3 panels 36,100 W POLYCRYSTALLINE CELLS with measures 2.5x66x101cm and 100 W of power in the standard STC test conditions for the off-grid system. For the calculation, the methodology of Pinho and Galdino [30] was used.

3.3 Daylighting analysis

The daylighting simulation reveals the availability of this resource in the coworking area, as shown in the Daylight Autonomy map (**Figure 11**). There is a predominance of natural light autonomy in the environment for over 80% of the hours during the year, with more than 300 lux. Illuminance values above 2000 lux, which can lead to glare and excessive thermal loads, are punctual and appear less than 40% of the time. In addition, they are concentrated exclusively along the building openings, as shown in **Figure 11**.

In general, and in terms of the high daylight availability when the environments are occupied, the results are satisfactory. Values with an autonomy of 300 lux less than 80% of the time are punctual (behind the wall and in rooms such as pantry and hallway, which usually do not have high lighting demand). Likewise, the compensation to reach higher levels, such as 500 lux in the work planes, can be contemplated by the work luminaires foreseen in the lighting project (task lighting). Furthermore, it is noteworthy that it would be highly restrictive to demand that the entire environment be served by 500 lux. In terms of potential glare, the 2000 lux Useful Daylight Illuminance analysis indicates dew occurrences near the windows, which can eventually be avoided

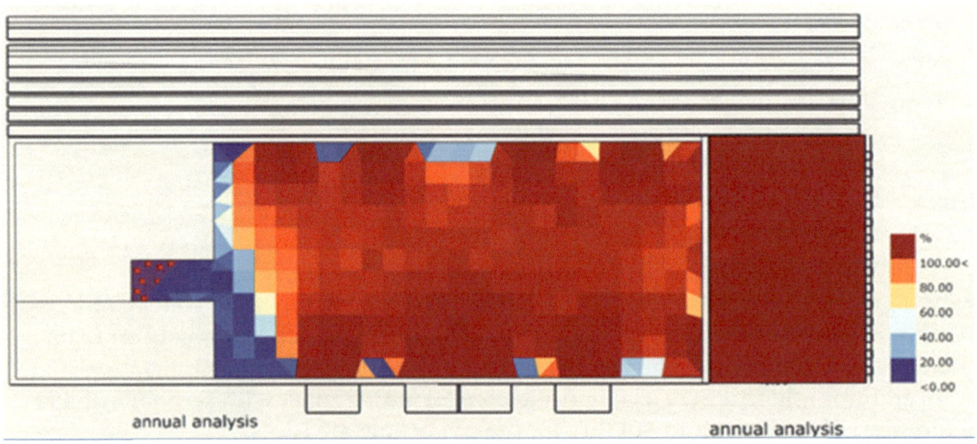

Figure 11.
300 lux daylight autonomy (DA) map for the Coworking and balcony area [25].

by adopting simple solar protection systems, such as blinds. In the external area and balcony, there are naturally higher rates, especially at the end of the building, which would probably benefit from some kind of greater protection (**Figure 12**).

3.4 Analysis of the lighting system

The adoption of high-efficiency solutions enabled an average illuminance of 411 lux in the coworking environment, as indicated by the simulations in Relux. At the workstations, the use of task luminaires that increase the illuminance to 500 lux on average is foreseen, as required by the NBR ISO/CIE 8995-1:2013 standard [31].

Thus, the project predicts a total of 54 luminaires, considering all areas and environments, with a total power of 801 W and a lighting power density (LPD) of 3.87 W/m^2. The minimum illuminance level required by NBR ISO/CIE 8995-1:2013 [31], entails an increase of 1 W/m^2, which raises the DPI to 4.87 W/m^2. Even so, this

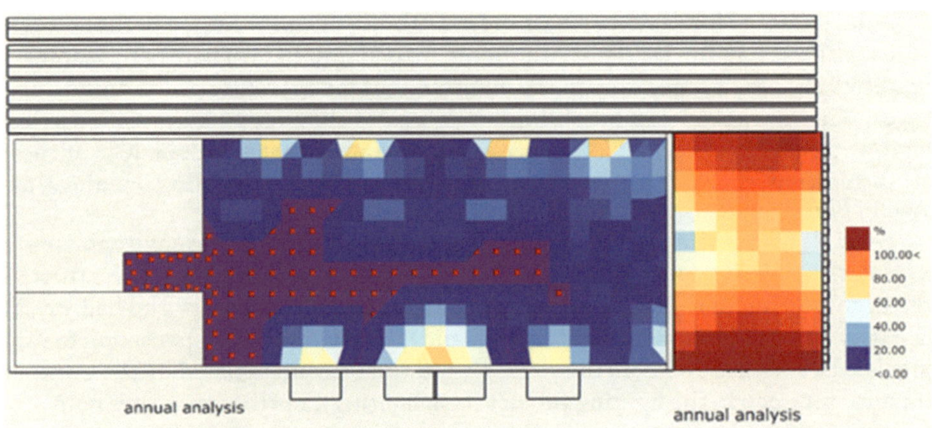

Figure 12.
Useful daylight Illuminance (UDI) map above 2000 lux for the Coworking and balcony area [25].

performance is considerably higher than the limit estimated by label A, according to the PBE Edifica PROCEL classification [11]. This demonstrates, in part, the potential for reducing LPD by using high-efficiency equipment.

This low LPD, combined with the control and automation system with sensors and dimming of the integration system between day and artificial lighting, allows a significant reduction in energy consumption. These elements are considered and verified later in the evaluation through simulation of energy performance.

3.5 Energy performance analysis

The building's energy consumption results assume a conservative scenario, with artificial conditioning of the coworking area throughout its occupation. However, the main objective of the project proposal foresees that conditioning should be applied only in situations when thermal comfort is not provided. Especially due to the great potential of using passive strategies. Nevertheless, it is prudent to take a conservative stance to ensure that the project will reach its goal of a building with a zero-energy balance.

Given the potential of taking advantage of natural light, the low demand for artificial lighting, the high-performance envelope, and efficient air conditioning equipment, it is possible to obtain an energy consumption of 34.30 kWh/m^2.year, as shown in **Table 6**. As a comparison criterion, the value obtained is considerably lower than the standards for corporate environments in Brasília-DF listed by [27], which demonstrates an average consumption of 131 kWh/m^2.year.

Thus, the division of consumption by final use, as shown in **Figure 13**, is considerably different from the typical consumption for commercial buildings foreseen by [32]. Unlike almost half (47%) of the energy consumption being related to the conditioning system, at LabZERO|UnB the conditioning system corresponds to 39%. It is worth noting that this reduction could be even more significant if a less conservative scenario were used regarding the air conditioning adoption. However, a greater reduction, from 22–12%, is seen in the artificial lighting system.

As for other electrical loads (equipment), demand exceeds the 31% predicted by [32], reaching 49% at LabZERO|UnB. This percentage does not reflect a quantitative increase in this type of load. However, it shows that its participation in the energy matrix of the building is greater. In part, this is justified by the fact that air conditioning and lighting systems are the main focus of these studies, being directly linked to architecture. For the calculation of energy demand of other electrical equipment, the standard was kept as a default. According to the very concept of building efficiency, the equipment adopted will probably follow the high-efficiency standards, which

End uses	Annual electrical energy consumption (KWH/year)	Annual electrical energy consumption (KWH/m^2. year)	Percentage (%)
Office Equipment	3451,13	16,67	49
Lighting	874,20	4,22	12
HVAC	2773,95	13,40	39
Total	7099,28	34,30	100

Table 6.
Consumption data by final and total uses per year and per year per square meter for the entire building [25].

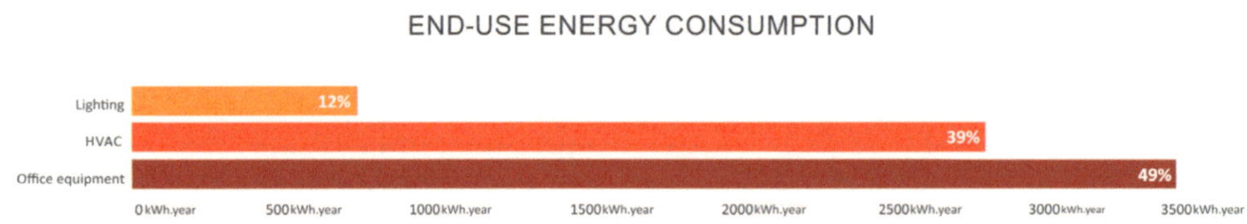

Figure 13.
Energy consumption by end-use [25].

reduces its demand. However, as the proposal aims to seek a more conservative scenario, this reduction was not considered for these environmental and energy performance analyses in the design stage.

3.6 Analysis of the potential for generation of photovoltaic energy

The graph in **Figure 14** shows the results of potential photovoltaic energy generation for the 2 systems (on-grid and off-grid), while **Table 7** presents the total values. It is observed that the on-grid system placed on the roof has significantly higher generation than the off-grid system, located in the brises, of 7,933 kWh/year and 4,155 kWh/year, respectively, which totals 12,088 kWh/year.

3.7 Building energy labeling

The building obtained a level A of energy efficiency label (the higher efficiency level, according to Brazilian National standard), as expected due to the inclusion of bioclimatic and energy efficiency strategies since its conception. Individually, the building envelope obtained EqNum = 5, the lighting system obtained EqNum DPI = 5, and the air conditioning system obtained EqNumVent = 5, related to the Coefficient of Performance (COP) of the machines. Considering a partial level A labeling

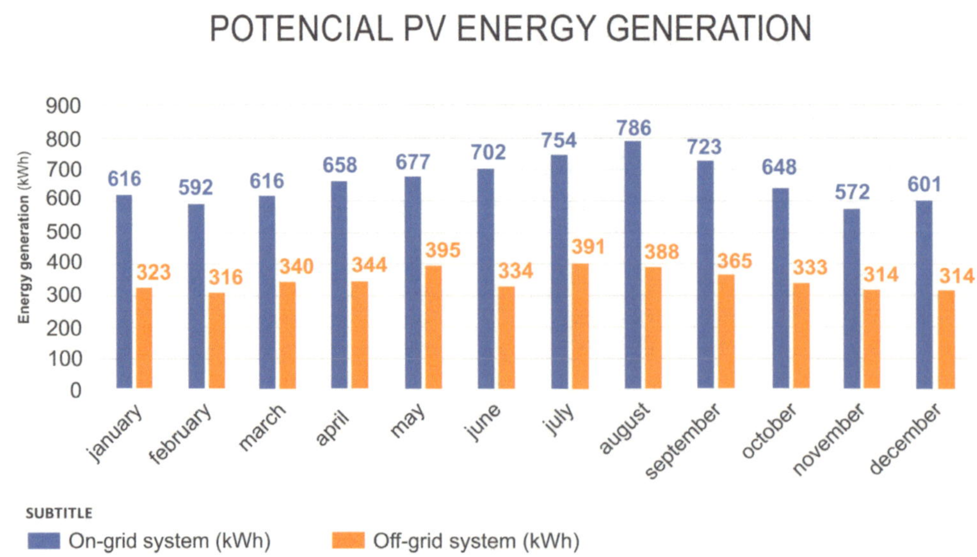

Figure 14.
Monthly estimated PV solar energy generation for LabZERO/UnB [25].

	On-grid (kWh)	Off-grid (kWh)	Total (kWh)
Power generation	7933.02	4155.18	12,088.2

Table 7.
Total values of photovoltaic energy generation potential in on-grid and off-grid systems.

obtained individually by these systems, plus a bonus for the rational use of water (40% savings), the energy savings from the network (more than 30%), and the general prerequisite of dividing electrical circuits fulfilled, the overall energy efficiency label obtained for the building is A, the most efficient.

3.8 Energy balance analysis

The graph in **Figure 15** shows the energy balance between consumption and generation. Even when considering the most conservative consumption, with the use of the conditioning system during the entire period of occupation, the simulations and final calculations prove that the building achieves an average annual electrical energy consumption of 34,29 kWh/m^2.year (7,099.18 kWh/year), which corresponds to a primary energy consumption value of 54,88 kWh/m^2.year (11,358.68 kWh/year). This is a significantly lower number if compared to the average consumption of electricity in office buildings in Brasília, which is close to 130 kWh/m^2.year [24]. As for the distributed generation of electricity in the photovoltaic system, installed on the roof and side area, the value of 58,29 kWh/m^2.year is obtained. The results are consistent with international experiences in similar climates, presented before (**Table 2**). With these data, the achievement of the building NZEB goals, or energy balance close to zero or nil, is achieved. There is, therefore, full compliance with the condition of the NZEB building (almost zero energy balance). It is also proposed that the energy generated in excess should be used to supply electric bikes and other buildings at the University of Brasília campus.

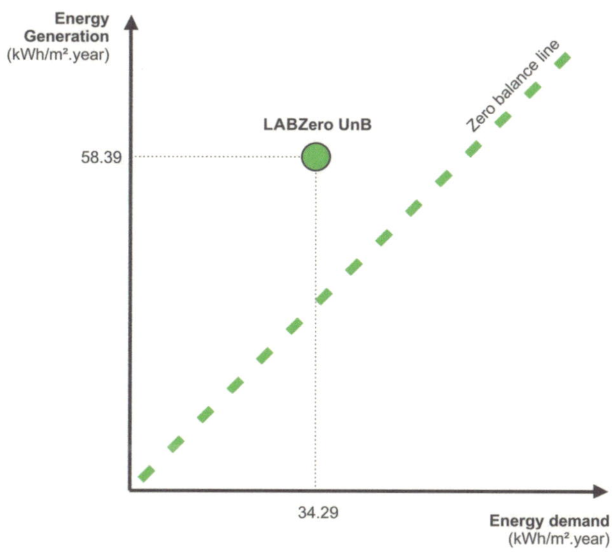

Figure 15.
Graph of the energy balance between building consumption and generation [25].

The building's reduced energy consumption is achieved through architectural and technological strategies (passive and active). In addition to aspects of energy efficiency and comfort, the building project proposes strategies for the rational use of water and waste management. Sustainability aspects are also highlighted. An example is the steel structure and the sealing in prefabricated panels, which allow for quick and clean construction, reducing waste generation and the replicability of the typology.

4. Conclusions and perspectives

The integrated design process, used as a methodology, proved to be efficient and highlighted the possibility of transposing research experiences into design practice. The barriers and potentialities related to the coordination of a multidisciplinary team and the organization, planning, and achievement of the goals in the integrated project process stand out. It is important to highlight the role of the computer simulation and the team in charge of this item in the design process, which must interact with others and effectively communicate the results. The project also underlines the importance of the facilitators, who coordinate the feedback loops of the computer simulations and architectural, the technological decisions between specialized teams and the group as a whole, in addition to defining deadlines and levels of detail for each specialty. Communication problems in the team can constitute barriers in the process, and the tendency of excessive detailing by experts at the beginning of the design process must be monitored by the facilitators.

The tools for analyzing environmental and energy performances through computer simulations are key parts to verify the zero-energy balance of a building and the fundamental elements in design decision-making. With these tools, the performance results can be accurately estimated.

In addition to being a building with a zero-energy balance, LabZERO|UnB is a project with the capability to achieve a positive energy balance, by offering an annual generation higher than its consumption. It has a demand of 34,29 kWh/m^2. year and a generation of 54, 88 kWh/m^2. year, which represents the potential to become a construction that has a positive energy balance with a 60% margin. This result occurs even considering conservative hypotheses of consumption reduction – such as constant use of artificial conditioning, with passive potential and office equipment with regular efficiency. Thus, the reduction in the energy consumption pattern from 131 kWh/m^2. year to 34 kWh/m^2. year is mainly due to solutions linked to the characteristics of the architectural project, such as shape, envelope, quantity, and opening orientation, combined with high-performance, artificial lighting, and mechanical conditioning systems. These indicate the advantage of considering environmental performance demands from the early stages of the project to achieve high-performance buildings.

It is expected that the construction of LabZERO|UnB, as well as the ELETROBRAS/PROCEL competition initiative, will be a milestone in the development of high-energy performance buildings in Brazil and zero-energy balance constructions. However, there is a need to incorporate environmental and energy performances analysis tools in the scope of the architectural project development from the preliminary stages, keeping in mind the operation and monitoring phases.

As a result, the project achieved an energy consumption of almost four times lower than the local average for office buildings, and this is compatible with international experiences. As the energy generation exceeds the demand, the NZEB building has the potential to supply other constructions or equipment. The strategies used for this combine the architecture plan conceived according to the local climate and directed towards the energy production in the building itself; and the main adoption of passive strategies, with the use of controlled active methods to optimize energy expenditure. After its construction, the building may be open to the public with a demonstrative purpose, allowing for large-scale dissemination.

Creating LabZERO|UnB reinforces the necessity of developing more sustainable and resilient buildings, with a possibility to extend the adopted strategies to other similar constructions creating, therefore, more efficient cities. This building will be a great model on the University campus, and it can be a prototype for future structures. It also works as a laboratory, in which people can better understand the importance of bioclimatic design and the incorporation of energy production on the building. Some architectural premises that were used on this project could also be applied in other constructions in the Brazilian context. Ultimately, LabZERO gives data to the Brazilian government to support public policies related to energy efficiency and sustainable energy production, all objectives which are bonded to the UN Sustainable Development Goals (SDG). In the context of the climate crisis, energy efficiency must be the natural strategy for developing countries in a tropical climate zone.

Acknowledgements

To the Infrastructure Secretariat and to the senior management of the University of Brasília, for their institutional support. Eletrobrás/Procel Edifica for supporting the project and CNPq for the Research Productivity grant.

Author details

Cláudia Naves David Amorim[1*], Joara Cronemberger Ribeiro Silva[1], Caio Frederico e Silva[1], Thiago Montenegro Góes[1], Ayana Dantas de Medeiros[1], João Manoel Dias Pimenta[2], Marco Antonio Egito[2], Adolfo Bauchspiess[2], Loana Nunes Velasco[3] and José Manoel Morales Sánchez[1]

1 Faculty of Architecture and Urbanism, University of Brasília, Brasília, Brazil

2 Faculty of Technology, University of Brasília, Brasília, Brazil

3 Faculty of Technology, Campus Gama, University of Brasília, Brazil

*Address all correspondence to: clamorim@unb.br

References

[1] Intergovernmental Panel on Climate Change – IPCC. Special Report. 48th Session. Incheon [Internet]. Geneva, Switzlerland: The Intergovernmental Panel on Climate Change (IPCC); 2018. Available from: https://www.ipcc.ch/site/assets/uploads/2019/07/SPM-Portuguese-version.pdf [Accessed: April 2, 2021]

[2] RUUSKA A, HÄKKINEN T. Material efficiency of building construction. Buildings. 2014;**4**:266-294. DOI: 10.3390/buildings4030266

[3] United Nations, Brazil. Transformando Nosso Mundo: A Agenda 2030 para o Desenvolvimento Sustentável. Brasilia, Brazil: Cúpula das Nações Unidas sobre o Desenvolvimento Sustentável [Internet]. 2016. Available from: https://www.br.undp-br-Agenda2030-completo-pt-br-2016.pdf [Accessed: April 2, 2021]

[4] United Nations, Brazil. Nova Agenda Urbana. Habitat III. Brasilia, Brazil: Quito [Internet]. 2016. Available from: https://habitat3.org/wp-content/uploads/NUA-Portuguese.pdf [Accessed: April 2, 2021]

[5] United Nations, Framework Convention on Climate Change. Adoption of the Paris Agreement. COP21, Paris. 2015. Available from: https://unfccc.int/documents/9064 [Accessed: April 2, 2021]

[6] TORCELLINI P, PLESS S, DERU M. Zero Energy Buildings: A Critical Look at the Definition. California, US: Proceedings of 2006 ACEEE Summer Study on Energy Efficiency in Buildings; 14–18 August 2006; 2006. pp. 3-428

[7] O'Brien W, Bourdoukan P, Delisle V, Yip A. Net ZEB design processes and tools. In: Athienitis A, O'Brien W, editors. Modelling, Design and Optimization of Net-Zero Energy Buildings. Berlin: Ernst & Sohn; 2015. pp. 107-174. DOI: 10.1002/9783433604625

[8] ASHRAE, AMERICAN SOCIETY OF HEATING. Advanced Energy Design Guide for Small to Medium Office Buildings: Achieving 50% Energy Savings Toward a Net Zero Energy Building. Atlanta, GA, USA: American Society of Heating, Refrigerating and Air-Conditioning Engineers; 2011

[9] EPBD RECAST. Directive 2010/31/EU of the European Parliament and of the Council of 19 May 2010 on the energy performance of buildings (recast) [S. l.: s. n.] [Internet]. Brussels, Belgium: Official Journal of the European Union; 2010. p. 28. Available from: https://eur-lex.europa.eu/legal-content/EN/TXT/?uri=CELEX%3A02010L0031-20210101 [Accessed: April 2, 2021]

[10] Keller M, Burke B. Fundamentos de projetos de edificações sustentáveis. Porto Alegre: Bookman; 2010 362 p

[11] BRASIL, RTQ-C: Regulamento Técnico da Qualidade para o Nível de Eficiência Energética de Edificações Comerciais, de Serviços e Públicos [Internet]. Brasília: PROCEL Edifica. 2010. Available from: http://www.pbeedifica.com.br/sites/default/files/projetos/etiquetagem/comercial/downloads/Port372-2010_RTQ_Def_Edificacoes-C_rev01.pdf [Accessed: April 4, 2021]

[12] ASSOCIAÇÃO BRASILEIRA DE NORMAS TÉCNICAS– ABNT. NBR 15575: 2021 - Edificações habitacionais - Desempenho. Parte 1: Requisitos Gerais. Rio de Janeiro: ABNT; 2021 98 p

[13] Agência Nacional de Energia Elétrica (ANEEL). Resolução Normativa N° 482, de 17 de abril de [Internet]. Brasilia, Brazil; 2012 Available from: http://www2.aneel.gov.br/cedoc/ren2012482.pdf [Accessed: April 2, 2021]

[14] Procel Edifica. Edital de Chamada Pública NZEB Brasil [Internet]. Rio de Janeiro, Brazil: Eletrobras; 2019. Available from: https://eletrobras.com/pt/Paginas/Chamada-Publica-Procel-Edifica2019.aspx

[15] Tauchen J, Brandli LL. A gestão ambiental em instituições de ensino superior: modelo para implantação em campus universitário. Gestão & Produção. 2006;**13**:503-515. DOI: 10.1590/S0104-530X2006000300012

[16] Kwok AG, Grondzik WT. The Green Studio Handbook. 1st ed. Oxford: Architectural Press (Elsevier); 2007. 378 p

[17] Monteiro LM, Bittencourt L, Yannas S. Arquitetura da Adaptação. In: Gonçalves J, Bode K, editors. Edifício. 1st ed. São Paulo: Oficina de Textos; 2015. pp. 27-53

[18] Mendes N, Amorim CND. Projeto Ambiental Integrado, uma experiência didática. In: Santos F, editor. Meio Ambiente em Foco. 12nd ed. Belo Horizonte: Atena; 2020. pp. 67-76. DOI: 10.36299/978-85-7042-215-6

[19] Garde F, Donn M. Solution sets and Net Zero Energy Buildings: A review of 30 Net ZEBs case studies worldwide. A technical report of Subtask C. IEA Task 40/Annex 52 Towards Net Zero Energy Solar Buildings. Le Tampon, Reunion, 2014. Available from: https://task40.iea-shc.org/Data/Sites/1/publications/T40A52-DC-TR1-30-Net-ZEBs.pdf [Accessed: March 2, 2021]

[20] Birck M. Projeto de edificação NZEB para o campus da UnB [thesis]. Brasília: Universidade de Brasília; 2014

[21] Pinto GSP, Pimenta J, Birck MB, Amorim CND. Proposta de edifício experimental com balanço energético nulo. Porto Alegre: Proceedings of 9th Congresso MERCOFRIO; 25-27 August 2014 Porto Alegre, Mercofrio; 2014. pp. 250-260

[22] Sudbrack LO. Casa Zero: diretrizes de projeto para casas pré-fabricadas de balanço energético nulo em Brasília [thesis]. Brasília: Universidade de Brasília; 2017

[23] ABNT—ASSOCIAÇÃO BRASILEIRA DE NORMAS TÉCNICAS. NBR 15220-3: Desempenho térmico de edificações – Parte 3: Zoneamento bioclimático brasileiro e estratégias de condicionamento térmico passivo para habitações de interesse social. Rio de Janeiro: ABNT; 2005

[24] CND A, Walter JFC, Sánchez JMM, Cronemberger J, Silva EG. Eficiência energética e iluminação natural: critérios para intervenção em edifícios modernos não residenciais em Brasília-DF. João Pessoa; Porto Alegre: XV Encontro Nacional de Conforto no Ambiente Construído e XI Encontro Latinoamericano de Conforto no Ambiente Construído (ENCAC/ELACAC); 18-21 September 2019, ANTAC; 2019. pp. 2314-2322

[25] AMORIM C, CRONEMBERGER J. Memorial Descritivo LabZero UnB: Proposta de Edificação de Balanço Energético Nulo. Brasília: Universidade de Brasília; 2020 76 p

[26] Towler G, Sinnott R. Chemical Engineering: Design Principles, Practice and Economics of Plant and Process

Design. 1st ed. Oxford: Butterworth-Heinemann (ELSEVIER); 2012 1320 p

[27] JFW C, OLIVEIRA NPR, CND AMORIM. Morphological characteristics and energy consumption of office buildings in the central area of Brasilia. Edinburgh: NCEUB: Proceedings of International Conference on Passive and Low Energy Architecture (PLEA); 3-5 July 2017; Edinburgh; 2017. pp. 2507-2514

[28] American Society of Heating, Refrigerating and Air-Conditioning Engineers. ASHRAE Handbook: Fundamentals. Atlanta, GA: ASHRAE; 2009

[29] Weber FS. Desenvolvimento de um modelo equivalente de avalição de propriedades térmicas para a elaboração de uma biblioteca de componentes construtivos brasileiros para o uso no programa EnergyPlus [thesis]. Florianópolis: Universidade Federal de Santa Catarina; 2018

[30] Grupo de Trabalho de Energia Solar – GTES (CEPEL – DTE - CRESESB). Pinho TT, Galdino MA, editors. Manual de Engenharia para Sistemas Fotovoltaicos. 1st ed. Rio de Janeiro: CEPEL – CRESEB; 2014 504 p

[31] ABNT—ASSOCIAÇÃO BRASILEIRA DE NORMAS TÉCNICAS. NBR ISO/CIE 9885-1: Iluminação de Ambientes de Trabalho – Parte 1: Interior. Rio de Janeiro: ABNT; 2013

[32] Eletrobras. Avaliação do mercado de eficiência energética no Brasil: sumário executivo – ano base 2005. Rio de Janeiro: Eletrobras; PROCEL; 2009. 187 p. Available from: [Accessed: March 3, 2021]

Chapter 6

Developing a Sustainable Solar-Residence Architecture Like a Home Unit without Energy Consumption from the Power Grid

Fernanda Antonio, Claudia Terezinha de Andrade Oliveira, Fabio Pires and Miguel Edgar Morales Udaeta

Abstract

The objective of this work is to identify the premises and strategies for the design of a zero-energy solar house and propose the systematization of its process. The focus of the application is on the single-family residential typology. The method consists of analyzing the whole process from the initial phase of the architecture project to the use of automation systems, aiming at the best use of solar energy in terms of sustainable development and high energy efficiency. Each phase of the process has significant importance in the performance of the residential unit, however, the influence that one phase has over another plays a fundamental role in the final result. The process of systematization encompasses all these phases, starting from the demands for energy in a solar house and introducing strategies to meet these demands. The prototype of the zero-energy solar house is used as an example of the application of this process for the development of a parametric solar house. The results show a strong positive correlation of linear dependence between the assumptions and strategies used in the architecture of the house and the solar system, allowing a conclusion of the dependence relation on sustainability, thermal comfort, visual and energy efficiency.

Keywords: zero-energy solar house, solar house, premises and strategies, sustainable development

1. Introduction

Methods of using and harnessing the sun's energy have been applied for a long time, providing several benefits to mankind. Growing global demand for energy and recent concern about the scarcity of natural resources have increased the demand for renewable energy, in particular, solar energy. However, the use of renewable energy is not the only way to achieve sustainable development. Increasing the efficiency of energy systems is also an effective way to reduce greenhouse gas (GHG) emissions and provide more energy for consumption.

In Brazil, hydroelectric power comprises 64.9% of the National Interconnected System (SIN), making the Brazilian electrical matrix comparable to those of developed countries. Nevertheless, the share of renewable energies can be further increased by the incorporation of solar systems in the residential sector, which would also impact national energy consumption. Studies show the huge potential for the exploitation of solar energy in the country, due to favorable levels of solar radiation throughout the year and because photovoltaic systems (PVSs) for distributed generation are approaching economic feasibility [1]. In addition, the residential sector accounts for 26% of total electricity consumption in the country, and it is expected that this participation will remain at the same level for the next 10 years, with an estimated increase of 48.3% by 2021 [2]. Such facts indicate the great impact that the incorporation of solar systems in the residential sector can cause on the national consumption of energy.

The objective of this study is to analyze and consolidate premises and guidelines for the design of a solar house. This project will focus on the development of a single family residential typology in terms of a zero-energy building (ZEB) aiming at sustainable development while maintaining environmental comfort and energy efficiency.

1.1 Current contribution on zero-energy solar house, with premises and strategies

The solar combisystem presents a high potential to contribute to the net-zero energy status of a building [3]. However, the energy requirement in a building is affected by parameters such as the envelope characteristics [4] and the weather [5]. The influence of these parameters on the energy demand can impact significantly energy management. Thereby, Raza et al. [6] proposed a demand forecast study of integrated photovoltaic intelligent buildings to improve their accuracy.

Another way to reduce energy consumption by the residential sector and increase the energy efficiency of solar systems is through the application of passive techniques to residential architecture. The incorporation of architectural features into residential buildings aiming at the full use of solar energy helps to maintain environmental comfort and reduce energy consumption. Studies on the bioclimatic architecture of buildings that use solar energy, either actively or passively, can be found in [7–10]. In this work, a housing unit that uses solar energy in architecture and energy generation, regardless of achieving a zero balance is defined as a 'solar house' [11].

According to Torcellini et al. [12], a ZEB is a building that produces, through local renewable sources, enough energy to equal or exceed its annual energy consumption. The ZEB can be connected to the public grid and integrates a system of distributed generation of electricity. In this study, the focus will be on the generation and use of solar energy by single family homes based on the development of a ZEB, regardless of reaching a zero-energy balance.

Initiatives to prospect, enable and analyze the quality of efficient solar energy projects were carried out in the USA, with competitions between universities to participate in the American Solar Decathlon Solar Energy (composed by Architecture, Engineering & Construction, Energy Efficiency, Communication & Social Awareness, Neighborhood Integration & Impact, Innovation & Viability, Circularity & Sustainability, Comfort Conditions, House Functioning and Energy Balance) in the years 2002, 2005, 2007, 2009, 2011, 2013, 2015, 2017, 2019 and then by the US and Spanish governments, which signed a memorandum of understanding in 2007 to create the first European edition complementary to the US Solar Energy Decathlon to encourage the use of solar energy and created the Solar Decathlon Europe. Spain hosted the first two competitions in 2010 and 2012 [13], the others were in France (2014), Hungary (2019) and the next in 2021/2022, in Germany [14].

Other regions were awarded the event, such as Solar Decathlon Latin America and Caribbean, in Colombia, which was held in 2015 and 2019, Solar Decathlon Africa, in Morocco (2016), Solar Decathlon China (2011, 2013, 2018, 2021), and Solar Decathlon India (2010/2021) [14]. Initiatives like these develop new sustainable technologies and transform society and the energy market.

Sustainable projects need to make several predictions, among others, of the air temperature in the almost zero energy building cooling system, with natural ventilation to reduce the dependence on mechanical systems and the use of models that can accurately predict the air temperature, therefore, in the naturally ventilated mode, they are fundamental to understand the current or future natural ventilation potential [15]. In southern European countries, summer temperatures can contribute to increased energy consumption, due to the cooling system of family units with less economic resources and may suffer from a lack of fuel to use air conditioning systems and one of the solutions to reduce the discomfort of thermal sensation is the optimization of natural ventilation to reduce energy consumption, improve cooling and serve social housing in coastal areas [16]. The European Union introduced the concept of construction of near-zero energy buildings in the system reformulation, using integrated solar technologies with dynamic occupancy in Finland and other northern European countries [17, 18]. Hybrid systems of integrated renewable energy, with a storage system, are important and efficient systems in the use of sustainable energy to serve localities [19].

In the study carried out by Zinzi and Mattoni [20], in the Italian cities of Rome and Turin, they described that it is important to guarantee energy efficiency, thermal comfort and indoor environmental quality, keeping construction and operation costs low. For the development of the project, three types of analysis were used: thermal comfort, energy performance and financial calculation. In both projects, the results were satisfactory.

The use of solar energy in buildings can mitigate the major challenges of energy shortages and global warming [21], and the management and integrated planning of energy resources are necessary to attract new investments, with the construction of sustainable environments, in projects that provide favorable conditions to the consumer market, with a reduction in the cost of electricity, with the application of innovative technologies in shared electricity generation, transmission and distribution and integrated [22].

To validate the study developed in this work, a solar house prototype used as a case study is taken as a study target. This prototype is a model of a zero-energy solar house (ZESH) and is analyzed qualitatively, considering general strategies for the passive use of solar energy. Furthermore, the ZESH concept is adapted to the Brazilian scenario where artificial lighting technologies, a domestic hot water system (DHW) and photovoltaic generations are considered for quantitative analysis.

Several works describe about zero energy solar house [12, 13, 15, 17, 19–21, 23, 24], but rare bibliography to the strategies used in zero energy solar house and this study seeks to complement the gap that exists on the subject, systematically contributing with new knowledge for decision making in the application of projects, using new premises and strategies, complementing the known strategies of innovation, sustainability and PVSs used in the design of a solar house (summarized in **Table 1**), as well as the basic principles for a zero energy solar house (ZESH). Fundamentally coming as a method for systematizing assumptions and strategies for the design of the mentioned zero-energy solar house, like several methods can influence the design results of a zero-energy solar

Strategy	Description
Solar geometry	It consists of the study of solar orientation, envelope volumetry, design and orientation of openings and sun protection elements. All these elements can be applied in the architectural design of buildings for optimizing comfort conditions and minimizing the use of energy consumption.
Thermal/visual comfort	The design of a bioclimatic architecture embraces aspects such as control of passive solar gains, natural ventilation and use of natural lighting. A. Controlling of passive solar gain can be achieved by selecting the building materials, the arrangement of the protection elements and the orientation of the building and openings. B. Natural ventilation and evaporative cooling are important to air renewal in the indoor environment, also this can be a strategic means of removing heat from the air. C. The combination of natural lighting and artificial lighting is an essential strategy used to meet the required demand for illumination in the internal environment of a building. The location and size of the openings and use of reflexive surfaces are important elements to maximize the use of natural lighting, ensuring visual comfort and minimizing energy consumption. For artificial lighting, the use of sensors, independent electric circuits and new technologies are some of the strategies used to control and enhance its efficiency.
Solar systems	It can be applied in the design of a building to maximize the use of solar radiation for water heating and electricity generation.
Residential automation systems	It can be applied for integration and monitoring the different systems operating in a solar house. Also, the information collected by these systems can be used to inform the occupants about the generation and consumption of energy to increase the energy efficiency consumption and comfort, ensure the safety of the occupants and maximize the operation of the systems.

Table 1.
Strategies for a solar house design.

house, among which are the solar geometry, the strategies used for thermal and/or visual comfort, as well as the solar house strategy, to consumption and installations.

2. Solar geometry

The position and angle of incidence of solar radiation on the facades of a building—throughout the day and year, for a particular site of implantation, considering the apparent movement of the sun—are determined by the study of solar geometry. This study can be used either for the use of solar radiation in a building, for example for natural lighting of the environment, or the protection of its surfaces from the direct incidence of the sun.

Buildings located in the tropics receive, in the summer, the less solar incidence in the north-facing facades (for the southern hemisphere) than east–west facing facades. However, in winter, the solar incidence of north-facing facades is higher than for those facing east–west. This indicates that geometries involving the use of more elongated facades orientated towards north and south obtain better use of the sun throughout the year.

Solar charts can contribute to the study of solar geometry. These charts graphically represent the apparent trajectory of the sun projected on a horizontal surface, for specific latitudes [25–27]. Their use allows to determine the hours of sunshine on horizontal and vertical surfaces for a given orientation [13]. In addition, it is also possible to determine, for a given day and time, the azimuth and solar elevation that can be used in the geometric study of the shading for the application of protection elements in facades [26].

The study of solar geometry and its application on architectural projects can also be carried out using physical and electronic models, assisting in the study of the effect of solar radiation and shadows on buildings. In this way, it is possible to adequately predict and design buildings in terms of solar orientation, envelope volumetry, design and orientation of openings and sun protection elements.

An accurate study considering all these aspects in architectural design can substantially influence the comfort conditions and the energy consumption for the environmental conditioning of a building.

3. Sustainable solar-house architecture

The air combines three parameters that influence the thermal comfort of an environment: temperature, humidity and speed (or movement). When climate conditions are analyzed based on temperature and humidity, it is possible to define four types of climates: hot and dry, hot and humid, cold and temperate [27, 28].

The architecture that seeks to adapt to the climatic conditions through its own design solutions and its own elements, favorably using these conditions and seeking to satisfy the thermal comfort needs of the occupants is a so-called bioclimatic architecture [27, 29, 30].

The study of the strategies to be adopted for each climate can be carried out using a bioclimatic (or psychrometric) chart. In these charts, through the relationship between temperature and relative air humidity, regions are defined in which it is possible to identify the most adequate strategies to reach the thermal comfort of the environment for a given location throughout the year.

The following items present the concepts related to the passive strategies oriented to the thermal comfort of the residential buildings and related to the control of the passive solar gains, natural ventilation and evaporative cooling.

3.1 Control of passive solar gain

Solar radiation can penetrate directly into the building through the openings or be absorbed by wall and roof surfaces. By focusing on the surfaces, solar radiation results in heat gain. This gain, and the consequent storage of heat in the environment, depends on the intensity of solar radiation and the thermal properties of both closing and internal materials to the building [25]. Direct gains occur when solar radiation strikes directly into the building, through lateral or zenithal openings. The use of transparent elements allows direct gains and can contribute to generate the greenhouse effect when there is a need for heating of the interior environment [30, 31]. Indirect gains occur when elements of high thermal inertia are exposed to direct solar radiation. These elements accumulate heat and emit it to the environment using radiation and convection, when the temperature of the internal environment decreases [25, 31]. **Figure 1** illustrates examples of passive forms of solar utilization that can be applied to the architecture of a solar house. Diagram (A) demonstrates direct or semi-indirect heat gain through a glasshouse. Diagrams (B), (C) and (D) show indirect passive gain forms where walls of high thermal mass accumulate the heat that will later be released to the environment to be heated.

However, it may be necessary to avoid solar gain in some climates or seasons of the year. Protection elements such as eaves, grilles, brises-soleil, marquises, external and

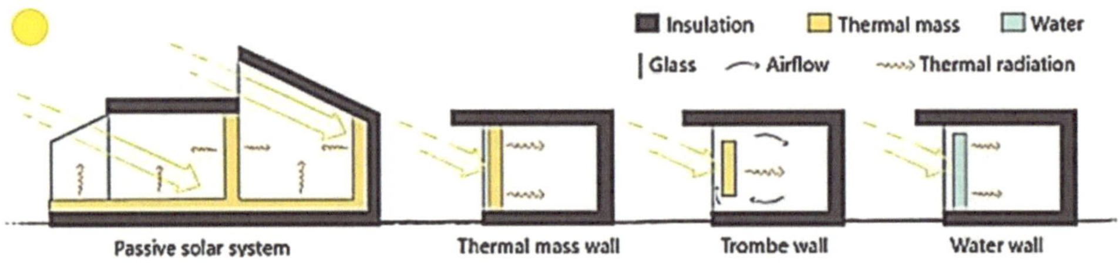

Figure 1.
Passive solar systems [21, 31].

internal blinds and vegetation can be applied together to the opening areas. The effect of solar protection of these elements on transparent surfaces depends on factors such as the ability of the material to reflect solar radiation, the location of this protection and its relative effect on solar radiation and heat convection, and their distribution about transparent surfaces [27].

Using selecting the building materials, the arrangement of the elements and the orientation of the building, these gains can be enhanced in the winter and inhibited in the summer. These solutions should be taken into consideration as these gains impact occupant comfort and energy efficiency to maintain environmental comfort.

The subsequent items 4-A1, 4-A2 and 4-A3 present features such as thermal inertia, thermal insulation and solar factor, respectively, which interfere in the transfer of heat, depending on the conditions of radiation incidence, opacity or transparency to radiation, surface conditions, mass, thickness and other materials properties and construction elements.

3.1.1 Thermal inertia

The thermal inertia is related to two important phenomena of the thermal behavior of buildings, the damping and the delay of the heat wave caused by the heating or cooling of materials. The greater the thermal inertia of the building, the greater the damping and the delay [25]. High thermal inertia materials can be used to control heat gains and losses in a building in the heat or cold [26].

The amplitude variation between the inner and outer temperature and the absorption of temperature peaks can be controlled by thermal inertia. Hot and dry climates are favorable to the application of this technique, due to the great thermal amplitude that they present between day and night. The thermal mass exposed to solar radiation absorbs heat during the day and returns it to the indoor environment at night when the temperature decreases. At night, the thermal mass cools and thus contributes to reducing the temperature of the internal environment during the day, since it absorbs heat again.

In cold climates, it is possible to use thermal mass in internal elements of the building, and thermal insulation in the external walls and openings, so that the heat retained in the material exposed to the solar radiation during the day is irradiated to the internal environment [31]. The insulation helps to preserve the heat in the interior environment, as well as the heat coming from appliances and the occupants' activities. Examples of high thermal mass materials are stones, massive bricks, adobe walls, concrete, water and green roofing.

The improper application of elements of high thermal inertia can act unfavorably to the internal environmental conditioning. Examples of this are the high thermal inertia buildings with mostly diurnal occupation or established in places where winter solar gains are insignificant. In these situations, the high thermal inertia may contribute to delay the reestablishment of the comfort conditions using air conditioning systems, increasing the energy consumption [32].

On the other hand, for low inertia buildings, Phase Change Materials (PCM) have been used to increase thermal mass [33]. Such materials absorb or release energy during their phase change, in temperature in the range of environmental comfort. Such a feature gives the material the ability to absorb temperature peaks. Examples of applications are the incorporation of PCM microcapsules into gypsum panels, into mortars or storage of material in tanks or compartments that are in contact with indoor air.

3.1.2 Thermal insulation

Thermal insulation is an indicated strategy for situations where thermal losses and gains are to be avoided. In cold climates, insulation is important for maintaining the heat inside the environment generated by passive solar gain strategies, or even active heating systems. In climates where temperature and humidity are high and artificial cooling systems are used, thermal insulation contributes to the use of these mechanical systems more efficiently.

The thermal insulation of a material can be determined by its coefficient of Thermal Conductivity (λ). The higher the value of λ, the lower the performance of the material as thermal insulation [30].

The choice of insulation depends on the need for insulation; performance in terms of durability, resistance and fire behavior; rainfall and technical issues regarding its installation in buildings.

3.1.3 Solar factor (SF)

The glass properties are an important factor in the energy fluxes through the openings, which can allow or inhibit thermal gains by solar radiation and heat losses of the interior environment [26, 30]. A characteristic commonly associated with glazed surfaces is the solar factor that is determined by the properties of absorptivity, emissivity and transmissivity. This factor represents the "relationship between the amount of energy flowing through a window and that strikes on it" [30].

An example of its use is in glasses with properties of low emissivity (or low-E). They reduce heat transmission by reducing the solar thermal gain, but allow the passage of visible light. It is, therefore, an alternative to allow natural lighting to contribute the control of heat exchanges.

3.2 Natural ventilation and evaporative cooling

Natural ventilation is essential to buildings because, in addition to promoting air renewal in the indoor environment, which is important for health, it contributes to thermal comfort in hot and humid climate regions and in hot periods, being one of the strategies indicated in the bioclimatic chart.

Natural ventilation can be described as the "displacement of air through the building, through openings, some functioning as an entrance and others as an outlet" [25]. Differences in air pressure between the internal and external environments influence this displacement, as well as the sizing and positioning of the openings. According to Frota & Schiffer [25], natural ventilation can be obtained in two different ways.

The first form occurs through the action of wind and is caused by the movement of air through the environment due to the force of the winds. The second form is obtained by the chimney effect, resulting from the difference in air density.

A strategy that can be adopted in conjunction with ventilation is evaporative cooling, which consists of removing heat from the air through the evaporation of water or evapotranspiration of plants. It is indicated for hot regions or periods with low relative humidity [30]. This strategy can be applied with the use of fountains, water mirrors and masses of vegetation in the vicinity of buildings. Water spray systems can also be adopted.

4. Strategies for visual comfort

Natural lighting influences the health and well-being of the occupants, for example by regulating the circadian system and mood, as well as contributing to the health of the environment, since sun exposure can eliminate viruses and bacteria.

Light affects the appearance of the indoor environment through general lighting and the performance of visual tasks that require specific levels of illumination. To meet the demand for lighting in the internal environment, natural and artificial lighting must be combined, aiming at the visual comfort of the occupants and the efficiency in the consumption of electric energy.

4.1 Natural lighting

In buildings, natural light can be obtained basically from three sources. The sun provides light through direct radiation, the sky provides diffused light, and the surfaces can provide reflected or indirect light [18]. The light of the sun and the light of the sky are both important, but they differ considerably in their characteristics. While sunlight results in high levels of illuminance, resulting in sharp contrast, diffuse light from the sky can result in more homogeneous illumination, with no excessive contrast between different points in the same environment, or between the interior and exterior of a building [34].

The Daylight Factor (DF) or Daylight Contribution (DC), is the ratio between the level of indoor and outdoor lighting [29, 30, 35]. This ratio is expressed as a percentage value. This percentage of lighting that will be available in the interior varies according to the size and location of the openings, the obstructions of the sky, the properties of the transparent closures to the light, and also according to the reflections of the interior surfaces [26].

According to ABNT [35] and Corbella [29], the luminous flux of the exterior can reach a point inside a building in three different ways. Using the light that reaches a point of the internal environment coming directly from the sky, defined as Sky Component (SC); by the light that reaches the interior after being reflected by external surfaces, defined as Externally Reflected Component (ERC) or by the light that is reflected by the surfaces of the internal environment itself, being determined by the shape, arrangement and colors of these surfaces, and defined as Internally Reflected Component (IRC).

Geometry also has an important influence on access to natural lighting inside buildings. Considering only the presence of openings in facades, distances of up to 5 meters can be illuminated naturally, and deeper distances are only partially illuminated [30]. Thus, for the same total constructed area, different geometries may allow greater or lesser access to natural light.

The illuminance in an environment increases with the size of the apertures. The Window-to-Wall Ratio (WWR) expresses the percentage value of the net area of an opening divided by the area of the wall containing it. Hastings, Wall [36] demonstrate in a study that, for WWR up to 50%, the increase of the illumination in the interior varies proportionally with the increase of the area of the opening. However, from a 50% WWR, the gains in terms of natural light are no longer so significant.

The use of windows on opposite facades of the same environment is a strategy that contributes to a more uniform distribution of natural light within buildings, reducing contrast and obfuscation. The orientation also influences the quality of natural lighting, as already mentioned in the Solar Geometry section. The graphs in **Figure 2** demonstrate the interior lighting curves for unilateral openings, and the illumination variation in the different orientations (the analysis in Phoenix is for reference only, to show the differences between Northern and Southern hemispheres).

For the use of natural lighting, different solutions can be adopted. The light shelves are horizontal devices that can be coupled to the windows, contributing to shade part of the opening, and allow greater penetration of the natural light inside the building, using the reflection of the light that in them penetrates to the ceiling of the internal environment, which favors the use of IRC [26, 29]. These elements can also avoid situations of discomfort and visual fatigue by contrast and dazzle [26].

Zenith lighting is a strategy that provides a more uniform distribution of lighting in the internal environment when compared to lateral lighting. Mansards, domes, sheds are examples of this type of application taking advantage of SC. Also, light tubes are elements capable of conducting natural light through a pipe into the building.

Sunlight can and should be used, however, strategies to take advantage of this feature must be appropriately integrated with other strategies to provide environmental comfort. Corbella [29] points out the importance of avoiding an excessive incidence of direct solar radiation in lighting projects in buildings located in tropical areas.

The sensitive use of natural light along with adequate elements to control the direct incidence of the sun and the levels of illumination in the interior should be considered an integral part of a passive solar project [34]. The single-family residential typology presents great flexibility to integrate different strategies and systems for the use of natural lighting, qualifying the internal environment and promoting visual comfort and efficiency in energy consumption.

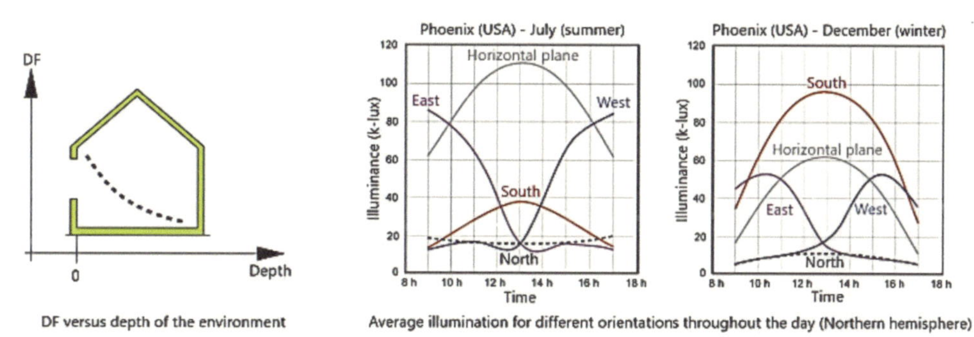

Figure 2.
Variation of natural lighting according to the depth of the environment and orientation of the openings [21].

Lamp Type	Watts	Lumens	Lifetime (hours)	Energy use (MJ/20 million lumen-hours)
Incandescent	60	900	1 k	15.1 k
CFL	15	900	8.5 k	3.78 k
LED	12.5–5.8	800	25–45 k	3.5–1.6 k

CFL: Compact fluorescent lamp; LED: light emitting diode; and MJ: Mili Joule.

Table 2.
Differences between artificial lighting systems.

4.2 Artificial lighting

A good artificial lighting project should be developed in a way complementary to the use of natural lighting available in each environment. The artificial lighting project comprises different steps, which include identifying the required lighting levels; the perception of the characteristics of the place to be illuminated, such as colors, types of surfaces and their dimensions; the choice of luminaires and lamps and their technical properties; to perform the calculation and determine the number of light points [29].

Among the strategies used in the control and efficiency of artificial lighting are the use of automatic systems such as photoelectric sensors, dimmers, presence sensors and time programmers and the adoption of independent electric circuits. An example of such circuits is the task lighting that is used to complement natural lighting and facilitate user autonomy by meeting their demand without having to provide higher illuminance throughout the environment.

The use of different artificial lighting technologies also has a great impact on energy consumption, the heat dissipated to the internal environment, the quality of the lighting and the need for system maintenance. Light Emitting Diodes (LED) systems have achieved market penetration and offer several advantages over incandescent and CFL lamps [37]. **Table 2** compares characteristics of LED, LFC and incandescent lamps.

Appropriate design strategies, adequate use of natural lighting and efficient technologies can contribute to the reduction of energy consumption by artificial lighting systems. In this way, it is possible to meet the demands of the occupants, satisfying the visual comfort parameters with energy efficiency.

5. Solar systems

The study of solar geometry is not limited to the use of the sun in passive strategies for environmental conditioning. It extends to the feasibility study of the use of solar systems, such as water heating and electricity generation. The analysis of the solar radiation in the different surfaces of the envelope of a building allows to take advantage of these systems. In new buildings, attentive to variation of solar radiation on surfaces allows to generate the shape of the envelope in order to obtain the best guidelines for the use of solar radiation in these systems, optimizing its yield.

5.1 Solar heating system (SHS)

Solar collectors are systems that convert sunlight, shortwave radiation that penetrates through the glass, into heat, and enable the use of solar radiation for heating water.

The SHS consists basically of three parts, the solar collectors, the thermal reservoir, and the water circulation system, which can be of natural circulation (thermosyphon) or forced circulation (pumped). In addition, they may also contain equipment such as a hydraulic pump and differential temperature controller.

The water heated by this system is stored in the thermal reservoir and will serve for consumption, supplying showers, sinks, kitchen sinks, appliances, and can also be used for heating the environment through, for example, water radiators.

The broadest SHS for residential use can be divided into three technologies, evacuated tube and glazed collectors, and unglazed collectors, the latter being the least efficient technology and whose application is mainly intended for pool heating [28]. **Figure 3** illustrates the difference in the performance of solar collectors.

The integration of SHS into buildings requires a study that considers the availability of envelope sites where these systems can be installed, the incidence of seasonal radiation on these surfaces, the design of the system based on hot water demand, and the space required for storage of heated water [38]. More efficient systems and better solar orientation result in a smaller area occupied by collectors for the same amount of heated water. The architectural design can also favor the efficiency of these systems by approaching points of collectors, storage and consumption, reducing losses with the transport of heated water.

From the economic point of view, Raimo [39] shows that the time of the return of the investment in SHS varies according to the solar coverage rate (SCR) and the efficiency of the system Thus, this period can be of a few months, for an efficient system and a high SCR, or reach about 8 years when the system presents low efficiency and the SCR is less favorable. The lifespan of these systems is about 20 years [31].

Another advantage of SHS is that the installation does not depend on specific legislation since these systems are not interconnected to public networks or infrastructures, such as the photovoltaic systems. On the other hand, the design of the system must be in accordance with the demand and also with the space available to store the volume of heated water. "Super or under-dimensioning of a solar heating system can turn out to be a worthless, unprofitable and even costly investment" [40].

Considering the sizing of the SHS, it is important to address the concept of Solar Fraction (SF), which is defined by the percentage of the total demand for hot water that

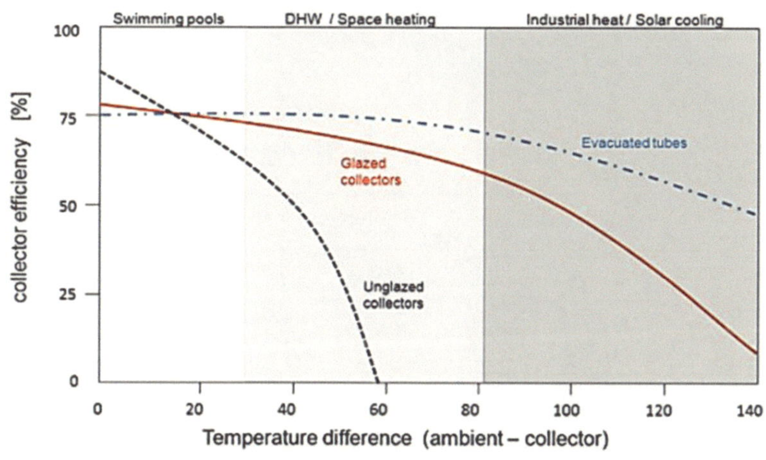

Figure 3.
Comparison between different technologies of solar collectors and their efficiency [38].

is supplied by the solar system [26, 31]. From the environmental point of view, the use of these systems has advantages in reducing GHG emissions when compared, for example, to electrical or gas systems. Avoided emissions can be measured based on the solar fraction that will have an SHS, comparing the energy that is no longer consumed from non-renewable sources, such as LPG, or even the electricity available in the network.

5.2 Photovoltaic system

Photovoltaic cells are independent power plants capable of converting solar energy directly into electrical energy. Several cells are interconnected to generate a photovoltaic module, which can be associated in series or parallel by configuring a larger power unit, and this association can be integrated with the buildings.

PVSs can be installed alone or connected to the distribution network. In Brazil, through a compensation system, a building connected to the grid can generate energy through a PVS. The surplus energy generated is injected into the grid and, when necessary, the building uses power supplied by the grid. PVSs connected to the grid do not require the use of batteries because they export surplus energy to the grid, which serves as a virtual storage system [38]. **Figure 4** shows schematically the components and operation of a photovoltaic system connected to the grid.

The integration of photovoltaic modules into the architecture can take different forms. Research focused on Building Integrated Photovoltaics (BIPV) have been contributing to the evolution and diversification of the photovoltaic modules available in the market, to facilitate this integration, in new and existing buildings. Examples of such applications are in windows and skylights, parapets and elements of roof or facade.

In addition, the integration of PVSs into building components contributes to the reduction of the final cost of these systems, since they are integrated with these components, instead of overlapping them, reducing costs with the module structure to photovoltaic cells [38]. **Figure 5** illustrates some of these possibilities.

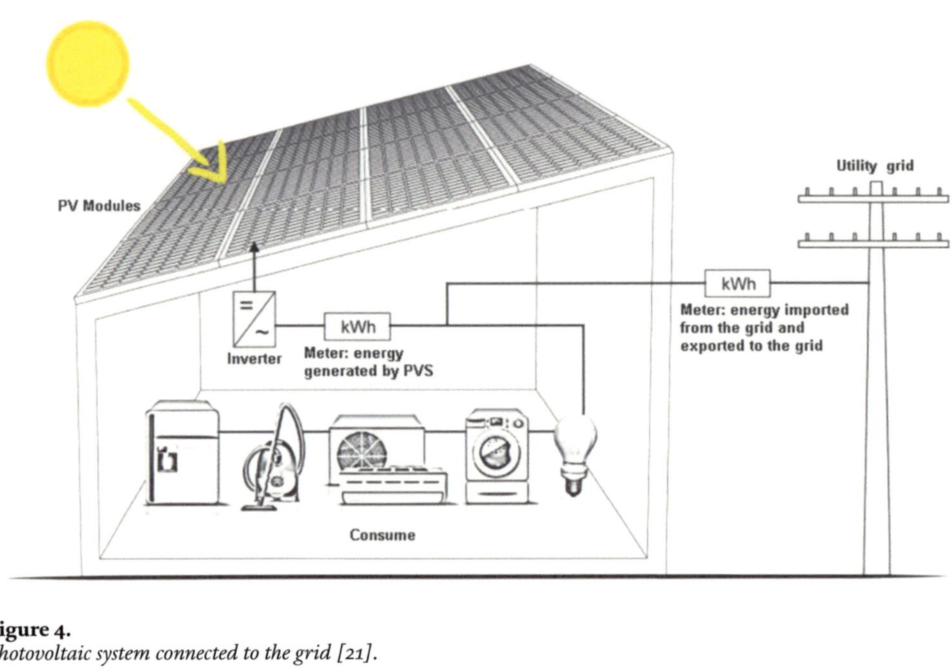

Figure 4.
Photovoltaic system connected to the grid [21].

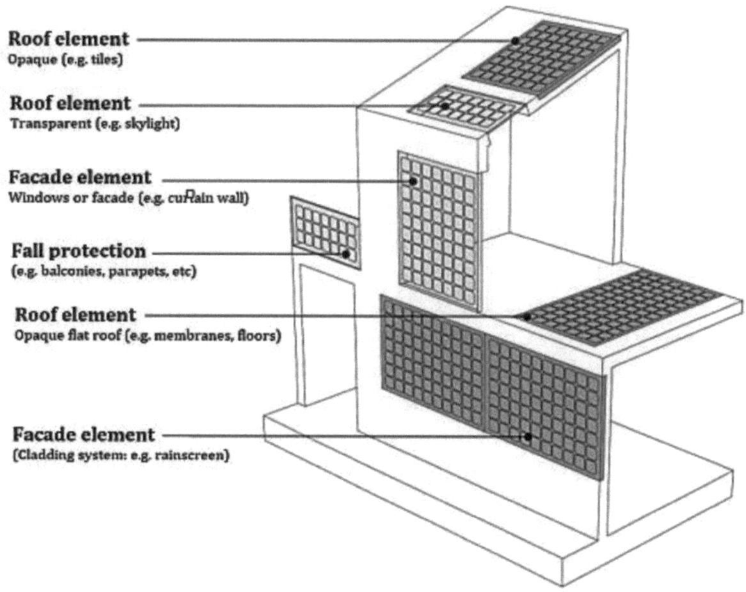

Figure 5.
Integration of PV modules to building components [21, 41].

From the environmental point of view, the PVS is a clean and renewable source of energy, not emitting gases like CO2, NOx and SO2 to generate electricity, and silicon modules are not toxic products, even at the manufacturing stage [31]. The lifespan of PVS consolidated technologies is 30 years, which the nominal power rating of the module is approximately 80% after 25 years of use, and inverters have a useful life of about 15 years [42]. A study carried out based on PVSs available in the market indicates that the time of return is less than 2 years for the main commercialized technologies [43].

Ruther [44] and Roaf et al. [31] point out some advantages of distributed photovoltaic generation in urban buildings, e.g. the reduction of losses in the transmission and distribution stages; the possibility of using the surfaces of the envelope for generation, without needing to occupy additional areas for the PVS; possibility of offering a high capacity factor to network feeders with daytime peak consumption; and modularity and ease of installation, providing quickly generation capacity to the distribution network. The absence of noise, reliability, low maintenance and the possibility of moving or transferring the system from one building to another, if necessary, are other positive aspects attributed to photovoltaic modules [31].

6. Energy consumption in a solar house

The electrical energy consumed in buildings is associated with the use of appliances and equipment, and these are generally the focus when it comes to reducing energy consumption. However, the consumption of these equipment depends not only on its efficiency, but also on the interaction with the envelope of the buildings and with the occupants [45]. Among the equipment used in a residence, those that have their demand and, therefore, their consumption more associated with the physical characteristics of the building are those used for the environmental conditioning, that is, artificial lighting systems and air conditioning.

To better understand residential energy consumption, let us take as an example the Brazilian case. Procel [46], on research into equipment checkout and use habits, rates the specific energy consumption of appliances and equipment in the Brazilian residential sector, as shown in **Figure 6**.

Bioclimatic strategies and the use of natural lighting can contribute to building performance, providing environmental comfort to the occupants and reducing the consumption of electricity with artificial lighting and air conditioning systems.

Decoupling of generation and consumption steps in the energy use cycle contributes to energy inefficiency. The ZEBs can reestablish this connection as they promote greater awareness of the occupants of energy generation and consumption. In this sense, residential automation systems are an important tool to integrate and monitor the different systems operating in a solar house, and inform the occupants about the generation and consumption of energy. The storage of the data of operation and performance of a house allows mapping tendencies that can be analyzed to look for solutions to increase the efficiency in the energy consumption for the maintenance of the environmental comfort [47].

Residential automation has been gaining market space as a way to increase not only the efficiency of the operation of the buildings, but also the comfort, convenience and safety of the occupants. The automation system integrates several components structured into a control skeleton that provides refined measurement data from sensors that detect equipment consumption, home appliances, electronics, lighting systems, power generation systems, temperature conditions, humidity, brightness, meteorological data, presence of people, among others. These, in turn, are registered and can be managed, or controlled, by users through interactive interfaces such as computers, mobile phones and the like.

Studies have shown the possibilities of avoiding waste of energy in energy consumption, or other natural resources such as water, through residential automation systems, which can be programmed to turn on or off equipment based on the presence of people, in the definition of temperature and lighting levels, helping to avoid wasted energy [47, 48].

Bartram and Woodbury [48] point out that the challenge of the automation project is to balance the responsibility of requesting actions on the part of the user and also

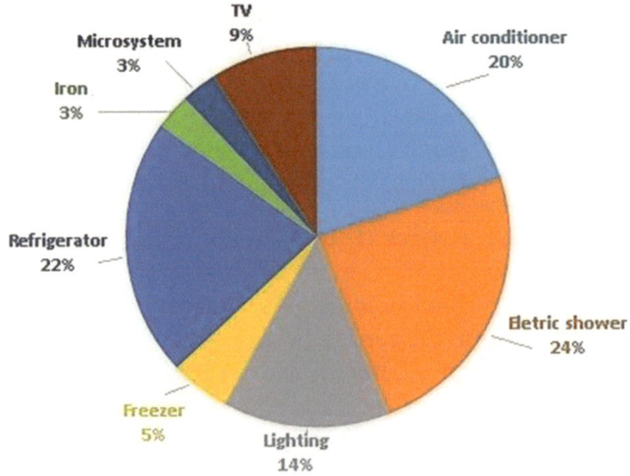

Figure 6.
Share of appliances consumption in Brazilian dwellings [46].

to assist in the accomplishment of these actions. The possibility of viewing historical expenditure data in monthly or annual periods tends to exert a great influence and impact on the user, which can also positively influence the seek for greater conservation of energy [44]. These systems and their interfaces present great potential to extend the design possibilities in homes that aim at energy efficiency and can instigate people to use natural resources more rationally.

6.1 Solar house model

The use of solar energy, directly or indirectly, in urban or rural buildings has great energy potential [38] and makes it possible to contemplate the energy demands to be used in a solar house. Examples of these uses are: allowing internal heating, which results from the generation of direct solar gains and also by the solar water heating system; effective use of natural lighting and use of shading elements to prevent internal overheating.

A solar house prototype is used as an example to show the application of different sun-use strategies in architecture and is the result of a study carried out in Madrid, Spain, located in the Northern Hemisphere, with geographic position 40.4168° N and 3.7038° W, where the implanted solar geometry was considered. The study allowed us to observe that the south orientation was considered the most favorable, as it benefits from the sun throughout the year. As a result, the geometry is more elongated on the east–west axis, while the largest area of openings is located on the south side of the prototype. However, the roof, which is the surface that receives the highest incidence of solar radiation throughout the year, was chosen for the installation of solar systems. The diagram in **Figure 7** illustrates these strategies [24].

The control of the incidence of solar radiation throughout the year is performed by shading devices, aiding in thermal comfort and the availability of natural light. In the southern facade, a system of automated external blinds was installed. The east and west facades are protected by verandas with bamboo frames. In the interior are applied translucent blinds. Components of high levels of thermal insulation are applied to the floor, walls and roof and the openings are sealed and double-glazed with a low-e coating. The application of these passive solutions for the maintenance of internal thermal comfort gave a daylight autonomy of 60%, contributing to the energy saving. The artificial lighting is designed

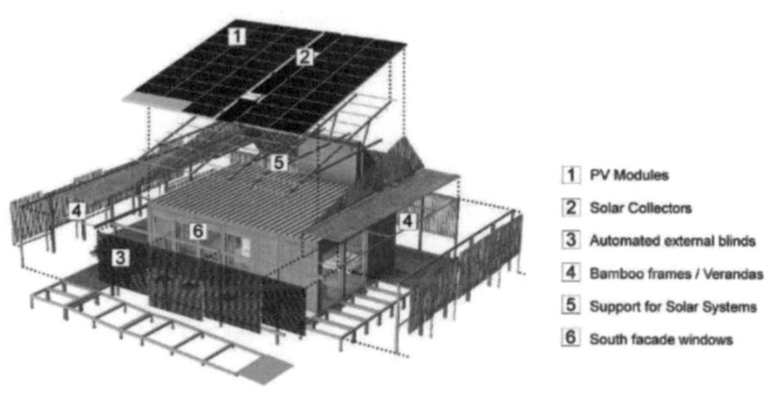

Figure 7.
Solar house prototype [24, 49].

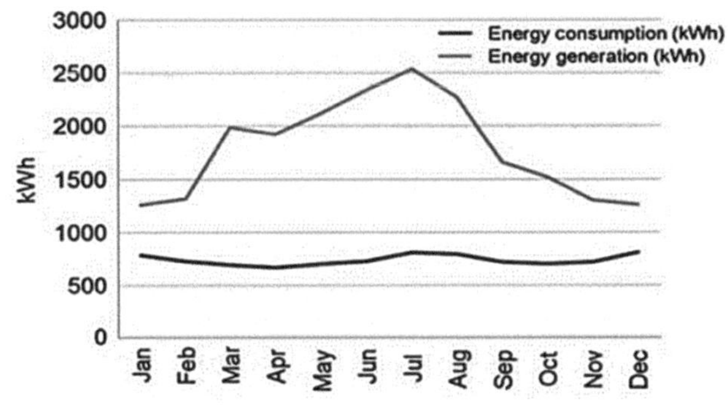

Figure 8.
Solar house annual energy balance considering the prototype located in Madrid [24, 49].

to complement the use of natural light and uses LED technology, which ensures greater efficiency than other technologies.

The application of a DHW solar system with evacuated pipe technology, with a solar fraction of about 90%, provides hot water, which can be used to feed radiators for space heating. The PV system consists of 48 modules with an efficiency level of 18.5%, accounting for an installed capacity of 11.04 kWp.

The efficiency of the operation is improved through the use of a home automation system integrated with the equipment and the general prototype process. The system provides to the occupant's information about power generation and consumption, allowing more efficient control of the use of appliances such as for lighting and thermal comfort. This equipment can be programmed to be activated or to work only under certain conditions pre-established by the occupants. The combination of solar systems and strategies for sun use guarantees the prototype a positive energy balance throughout the year, as shown in the graph of **Figure 8**.

The estimation of the annual energy balance of the House prototype was carried out considering its implementation in Madrid. According to the rules and regulations of the SDE, an occupancy schedule was defined and energy consumption of household appliances was estimated considering a couple's daily routine and interior comfort conditions for certain temperature and lighting ranges. Energy consumption and generation calculations, conducted using Energy Plus software by Team Brazil members, also took into account the climate data from Madrid. For the calculation of energy generation, was adopted the same PV panel used in the prototype, which is a monocrystalline technology SunPower 230 Solar Panel with a 15.8% efficiency, and the solar collectors were vacuum system SOLTER PU 200/5 [21].

The prototype Solar House adopts as premise the harnessing of sun. It exemplifies a model of the solar house, designed in the light of an adequate study of solar geometry and solar orientation. In addition, it shows that the combined use of strategies and systems can improve the performance and efficiency of the housing unit.

7. Systematization of premises and strategies for a solar house

Distinct is the premises and strategies for using the sun in the architecture of a solar house, be they design solutions or systems that may be incorporated into the building. As described throughout this study, and demonstrated through the Solar

House prototype, these strategies relate to one another, interfering with each other, as well as with the outcome of the edification as a whole. Therefore, the earlier the architectural use of the sun is taken into account in the design process, as a guideline in the choices of premises and the strategies to adopt, the greater the benefits arising from the use of this resource in the architecture.

In addition, many factors may limit the application of these assumptions and strategies, e.g., economic, cultural, technical, technological, or other. It is important to see in a systemic way the possible strategies to be adopted and the interfaces between them, to obtain greater energy efficiency and environmental quality as a result, within the limitations of each project. The diagram shown in **Figure 9** starts from the use of solar energy as a fundamental premise in the design of a Solar House. Then the energy demands of a solar house are incorporated and, sequentially, the strategies that can be applied to the use of the sun in the architecture are added to the diagram, increasing the use of technologies and the consequent complexity of the project.

In this work, the premises and strategies considered the most relevant within the scope of this research were listed. There is a multitude of other strategies, or even derivations of the demands and strategies presented. In this way, it would be possible to incorporate to this diagram structure new elements, increasing its complexity and refinement regarding the adoption of design solutions for a solar house.

Thus, starting from the use of the sun as a premise, a housing unit can be considered the most elementary version of a solar house when conceived considering an adequate study of geometry and solar orientation. Other strategies and systems can be incorporated, through different arrangements and combinations of these solutions, improving the performance and efficiency of this unit. In addition, preparing the building so that strategies can be incorporated into future steps, for example, leaving waits for SHS and PVS when they cannot be adopted at first, is also essential considering the lifespan of a solar house.

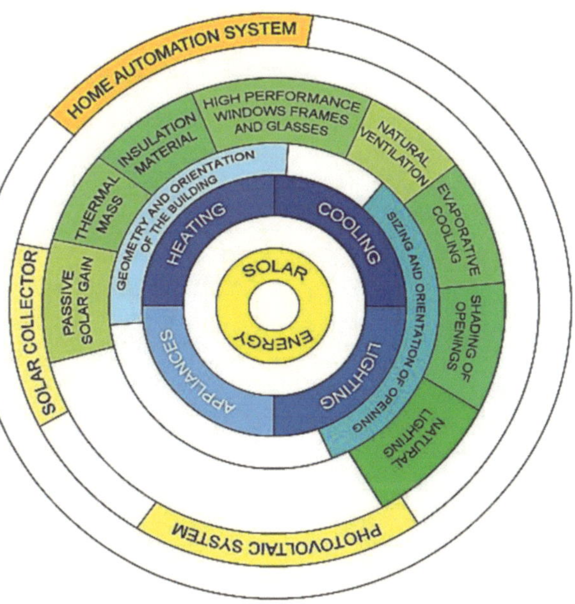

Figure 9.
Demands, premises and design strategies for a solar house.

Table 1 presents the summaries of the assumptions and strategies of solar geometry, thermal and visual comfort, solar systems and home automation.

8. Outcomes analyses and final considerations

The use of the premises and strategies of the solar house project allowed satisfactory results in the use of architectural techniques and technologies to keep the climate control and the visual lighting system comfortable [23], as well as resulted in the generation of electricity, with gains physical and economic resources applied to the project.

The Solar House project was dimensioned with balconies and designs to control heat exchanges, for thermal comfort and with large openings to amplify the use of natural light, to aid visual comfort. The automation system, with controllers designed to reduce energy consumption as a function of demand [50, 51] allows the interaction between the resources of the solar system, such as thermal heating, lighting and energy generation by photovoltaic cells [52] and equipment for general use, controlling electricity consumption.

In addition to individual and/or collective projects in the implementation of solar systems, with highly satisfactory results, some countries, such as Denmark, China, Germany and Austria, have made extensive investments in the energy market, with technological solutions in large-scale solar thermal systems [53]. Several contributions from solar resources have motivated the implementation of these models of thermal and electrical energy generation, as they are non-polluting and provide a clean form of energy. In **Figure 10**, the diagram systematically presents the results found.

In Section 6, **Figure 8**, the excess energy available for the electricity grid is presented, at the approximate average value of 1030 kWh/month, equivalent to 58.79% of the total and **Table 3** [21] reinforces the positive results in implantation of solar systems, including the simulated average values of energy generation and consumption in one year.

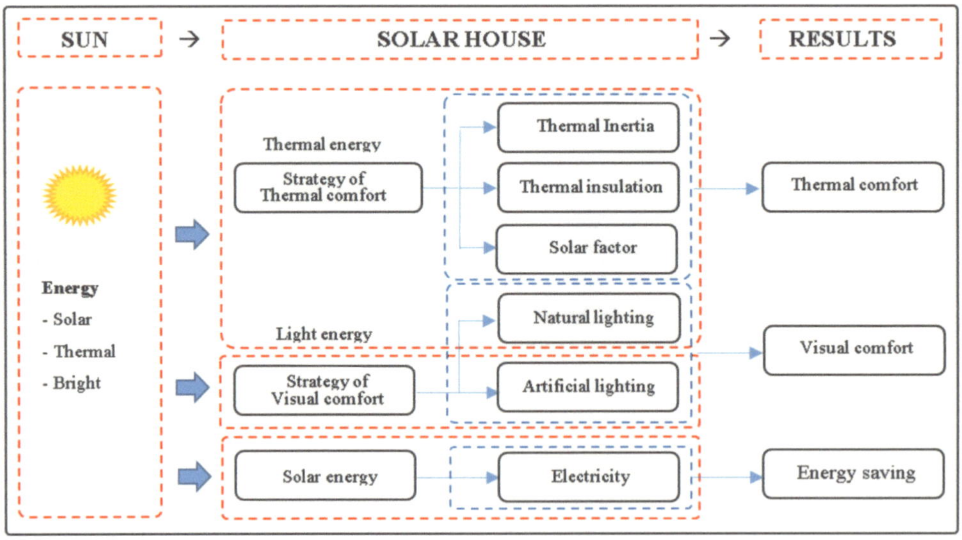

Figure 10.
Diagram of solar energy for a solar house.

DOI: http://dx.doi.org/10.5772/intechopen.102778

Month	Generated energy (KWh)	Energy consumed (KWh)	Energy exceeded (KWh)	Percentage (%)
Jan	1253	747	506	40.38
Feb	1387	721	666	48.02
Mar	1852	679	1173	63.34
Apr	1918	634	1284	66.94
May	2204	676	1528	69.33
Jun	2405	743	1662	69.11
Jul	2451	801	1650	67.32
Aug	2012	785	1227	60.98
Sep	1599	710	889	55.60
Oct	1496	688	808	54.01
Nov	1268	728	540	42.59
Dec	1182	754	428	36.21
Average	1752.25	722.17	1030.08	58.79

KWh: KiloWatt-hour.

Table 3.
Generated energy capacity (KWh).

The data allow showing the Pearson correlation, with r = 0.995, demonstrating a strong positive relationship between generated energy and surplus energy (energy produced minus consumed), which returns to the electricity grid. The percentage of energy exceeded, 58.79%, demonstrates the efficiency level of the solar house's energy system.

The experience in the solar house allowed us to verify some advantages and disadvantages in its implementation in relation to the materials, techniques and equipment of the solar system:

a. Advantages—allows you to model hourly variations in occupancy, simulate bioclimatic strategies, decrease the cost of electricity, due to the improvement in thermal comfort (design of windows, doors, side openings, thermal plates, natural ventilation network, artificial conditioning, thermal control, etc.) and visual (natural and artificial lighting system).

b. Disadvantages—high initial cost for installing the solar house, due to materials, equipment, as well as the limited energy generation during the day.

The parameters and variables allow achieving thermal, visual, acoustic, air quality, etc. comfort, depending on the architecture, solar geometry, air temperature, radiant temperature, relative humidity, air velocity, physical activity, clothing, thermal exchanges: conduction, convection, irradiation (thermal gain), etc. They are important to the process of analyzing the premises and strategies of a solar house.

The solar intensity, as well as the temperatures and relative humidity of the air, are important variables for the design of the solar house and, in this sense, the averages, maximum and minimum values of the cities of Madrid and Sao Paulo are presented.

Average temperatures (°C) and relative humidity (%):

Madrid—Average temperature range: from 0–33°C (Min −9°C and Max. 40°C), with average annual temperature: 14°C. Relative air humidity: 10 to 80%, with average annual humidity: 55.4% [54].

Sao Paulo—Average temperature range: from 12–28°C (Min 2°C and Max. 40.4°C), with average annual temperature: 26.1°C. Relative air humidity: 12 to 88%, with average annual humidity: 79.56% [55].

Table 4 supports the proposal of the efficiency of the solar house, given the growth of the solar energy matrix in Brazil. The Brazilian installed power grew approximately 130% from 2018 to 2019, 70% from 2019 to 2020 and in the first four months of 2021, it had more than 15% [56]. These data provide the potential of the Brazilian electrical matrix about the generation of solar energy, which has a large installation capacity in the country, due to the climate, the reduction of implementation costs and, mainly, for the sustainability and improvement of climate comfort, visual and environmental.

Year	Solar energy Brazil		Electric generation capacity	
	Installed capacity (GW)	Growth (%)	Overall capacity (GW)	Ratio (solar energy)/(general energy) (%)
2016	0.024	14.29	150.338	0.016
2017	0.935	3795.83	157.112	0.595
2018	1.798	92.30	162.840	1.104
2019	2.473	37.54	170.118	1.454
2020	3.287	32.92	174.412	1.885

GW: Giga Watt.

Table 4.
Solar energy capacity in Brazil [56].

	Solar power	General power
Average (GW)	1.703	162.964
Variance (GW)	1.630	93.978
Standard deviation (GW)	1.277	9.694
Comments	5	5
Correlation (r)	0.963	
Mean difference hypothesis	0	
gl	4	
Stat t	−36.87782931	
P(T<=t) one-tail	1.61411E-06	
one-tailed critical t	2.131846786	
P(T<=t) two-tailed	3.22823E-06	
two-tailed critical t	2.776445105	

GW: Giga Watt; r: Pearson correlation; gl: degree of freedom; Stat t: t statistic; P: Probability.

Table 5.
Correlation of statistical data on Brazilian solar and energy capacity.

The installed solar power in Brazil, in 2018, was 1.8 GW and corresponded to 1.1%, in 2019 it increased to 2.47 GW and, in 2020, to 3.29 GW, representing 1.88% of the matrix Brazilian energy, therefore, shows a significant growth of solar energy in the country.

Table 5 presents the statistical data, demonstrating a strong relationship between the growth of the capacity of the solar system and the Brazilian energy system, as well as the significance and importance of this modal for the Brazilian energy system.

Strategies and premises are important for the solar house process and statistical data prove this, with r = 0.963 and P-value <5% (0.00000161).

9. Conclusions

In this work, several solutions were presented through which the architecture can benefit from the sun to provide environmental comfort to the occupants and efficiency in the consumption of electricity. The single-family housing typology, due to its characteristics, presents an ease to incorporate many of the premises and strategies presented, from those involving only the study of solar geometry and orientation in relation to the sun, to the incorporation of sophisticated systems and technologies.

The theoretical approach and the practical examples of bioclimatic concepts, material properties, design strategies, solar systems, equipment and technologies for energy conservation, allow us to intuitively understand that all these solutions, if well used, contribute to an architecture of quality, for the benefit of the occupants and the environment. The prototype Solar House was used to demonstrate the practical application of these premises and strategies, as well as the interfaces between them.

The results demonstrate the importance of the solar system for the Brazilian energy matrix and how solar houses contribute to this process of energy reduction, through the sustainable solutions presented in this study.

Finally, it is pointed out the contribution of this type of housing unit towards sustainable development, leading to a reduction in greenhouse gas emissions [24, 49].

Acknowledgements

To Daniela Miwa Uemura (Msc) for the translation from Portuguese to English of this work in the first version, with the content and form, and to Prof. André Luiz Lorenção for general review.

Author details

Fernanda Antonio[1], Claudia Terezinha de Andrade Oliveira[1], Fabio Pires[2] and Miguel Edgar Morales Udaeta[2]*

1 Faculdade de Arquitetura e Urbanismo da USP (FAU/USP), São Paulo, Brazil

2 Grupo de Energia do Departamento de Engenharia de Energia e Automação Elétricas da Escola Politecnica da USP (GEPEA/EPUSP), São Paulo, Brazil

*Address all correspondence to: udaeta@pea.usp.br

References

[1] MME, EPE. Balanço Energético Brasileiro: Relatório Síntese 2021. Brasília: Ministério de Minas e Energia, Empresa de Pesquisa Energética; 2021

[2] MME, EPE. Plano Decenal de Expansão de Energia 2021. Brasília: Ministério de Minas e Energia, Empresa de Pesquisa Energética; 2012

[3] Asaee SR, Ugursal VI, Beausoleil-Morrison I, Ben-Abdallah N. Preliminary study for solar combisystem potential in Canadian houses. Applied Energy. 2014;**510**:518-130. DOI: 10.1016/j.apenergy.2013.12.048

[4] Ascione F, Masi RFD, Rossi F, Ruggiero S, Vanoli GP. Optimization of building envelope design for nZEBs in Mediterranean climate: Performance analysis of residential case study. Applied Energy. 2016;**938**:957-183. DOI: 10.1016/j.apenergy.2016.09.027

[5] Fikru MG, Gautier L. The impact of weather variation on energy consumption in residential houses. Applied Energy. 2015;**19**:30-144. DOI: 10.1016/j.apenergy.2015.01.040

[6] Raza MQ, Nadarajah M, Ekanayake C. Demand forecast of PV integrated bioclimatic buildings using ensemble framework. Applied Energy. 2017;**1626**: 1638-1208. DOI: 10.1016/j.apenergy.2017.08.192

[7] Chwieduk. Towards sustainable-energy buildings. Applied Energy. 2003;**211**:217-276. DOI: 10.1016/S0306-2619(03)00059-X

[8] Ahmed AZ, Sopian K, Othman MYH, Sayigh AAM, Surendran. Daylighting as a passive solar design strategy in tropical buildings: A case study of Malaysia. Energy Conversion and Management. 2002;**1725**:1736-1743. DOI: 10.1016/S0196-8904(01)00007-3

[9] Llovera J, Potau X, Medrano M, Cabeza LF. Design and performance of energy-efficient solar residential house in Andorra. Applied Energy. 2011;**1343**:1353-1388. DOI: 10.1016/j.apenergy.2010.10.015

[10] Huang KT, Hwang RL. Future trends of residential building cooling energy and passive adaptation measures to counteract climate change: The case of Taiwan. Applied Energy. 2016;**1230**:1240-1184. DOI: 10.1016/j.apenergy.2015.11.008

[11] Antonio F. Premissas e estrategias para uma casa solar visando a` redução de emissões de gases de efeito estufa por meio da conservação e geração de energia [dissertation]. São Paulo: Universidade de São Paulo; 2015

[12] Torcellini P, Pless S, Deru M. Zero energy buildings: a critical look at the definition. In: ACEEE Summer Study. California: Pacific Grave (USA) NREL; 2006 Conference Paper NREL/CP550-39833

[13] Yu Z, Gou Z, Qian F, Fu J, Tao Y. Towards an optimized zero energy solar house: A critical analysis of passive and active design strategies used in Solar Decathlon Europe in Madrid. Journal of Cleaner Production. 2019;**236**:117646. DOI: 10.1016/j.jclepro.2019.117646

[14] US (SD). Solar Decathlon Solar de Energia nos EUA, Europa, China, India, Africa, América latina e Caribe, Americano 2021. Available from: https://www.solardecathlon.gov/international-europe.html

[15] O'Donovan A, O'Sullivan PD, Murphy MD. Predicting air temperatures

in a naturally ventilated nearly zero energy building: Calibration, validation, analysis and approaches. Applied Energy. 2019;**250**:991-1010. DOI: 10.1016/j.apenergy.2019.04.082

[16] Bienvenido-Huertas D, Sánchez-García D, Rubio-Bellido C. Analysing natural ventilation to reduce the cooling energy consumption and the fuel poverty of social dwellings in coastal zones. Applied Energy. 2020;**279**:115845. DOI: 10.1016/j.apenergy.2020.115845

[17] Reda F, Fatima Z. Northern European nearly zero energy building concepts for apartment buildings using integrated solar technologies and dynamic occupancy profile: Focus on Finland and other Northern European countries. Applied Energy. 2019;**237**:598-617. DOI: 10.1016/j.apenergy.2019.01.029

[18] Colclough S, McGrath T. Net energy analysis of a solar combi system with Seasonal Thermal Energy Store. Applied Energy. 2015;**147**:611-616. DOI: 10.1016/j.apenergy.2015.02.088

[19] Liu J, Chen X, Yang H, Shan K. Hybrid renewable energy applications in zero-energy buildings and communities integrating battery and hydrogen vehicle storage. Applied Energy. 2021;**290**:116733. DOI: 10.1016/j.apenergy.2021.116733

[20] Zinzi M, Mattoni B. Assessment of construction cost reduction of nearly zero energy dwellings in a life cycle perspective. Applied Energy. 2019;**251**:113326. DOI: 10.1016/j.apenergy.2019.113326

[21] Ren H, Sun Y, Albdoor AK, Tyagi VV, Pandey AK, Ma Z. Improving energy flexibility of a net-zero energy house using a solar-assisted air conditioning system with thermal energy storage and demand-side management. Applied Energy. 2021;**285**:116433. DOI: 10.1016/j.apenergy.2021.116433

[22] Pires F, Udaeta MEM, Dias EM, Pires FS. Challenges of Integrated Planning, both for supply and demand, in the contingency of energy policies. In: XI CBPE—Congresso Brasileiro de Planejamento Energético. Vol. 2018. Cuiabá: Congresso Brasileiro de Planejamento Energético; 2018

[23] El Sayary S, Omar O. Designing a BIM energy-consumption template to calculate and achieve a net-zero-energy house. Solar Energy. 2021;**216**(1):315-320. DOI: 10.1016/j.solener.2021.01.003

[24] Oliveira CT, Antonio F, Burani GF, Udaeta MEM. GHG reduction and energy efficiency analyses in a zero-energy solar house archetype. International Journal of Low-Carbon Technologies. 2017;**225**:232-212. DOI: 10.1093/ijlct/ctx004

[25] Frota AB, Schiffer SR. Manual de Conforto térmico. 8th ed. São Paulo: Studio Nobel; 2003

[26] Brown GZ, Dekay M. Sol, Vento e Luz: estratégias Para o Projeto arquitetônico. Porto Alegre: Bookman; 2004

[27] Olgyay V. Arquitectura y Clima: "Manual de Diseno bioclimático Para Arquitectos y Urbanistas". Barcelona: Gustavo Gili; 1998

[28] Serra R. Arquitectura y Climas. Barcelona: Gustavo Gili; 1999

[29] Corbella O, Yannas S. Em Busca de uma Arquitetura sustentável Para os trópicos: Conforto Ambiental. Rio de Janeiro: Revan; 2003

[30] Lamberts R, Dutra L, Pereira FOR. Eficiência energética na arquitetura. Rio de Janeiro: ELETROBRAS, PROCEL Programa Nacional de Conservação de Energia Elétrica, PROCEL EDIFICA Eficiência energética em Edificações. 2014

[31] Roaf S. Ecohouse: A Design Guide. Arquitectural Press; 2007

[32] LIMA MAS. Armazenamento de Energia térmica Em Componentes de edifícios Sobre os Quais Incide a radiação Solar Directa [Dissertation]. Porto: Universidade do Porto; 1995

[33] Santos ACN. Estudo da eficiência energética de Um Sistema de Arrefecimento Com Materiais de mudança de Fase [Dissertation]. Porto: Universidade do Porto; 2013

[34] BSI. Quality management systems. Requirements. In: BS EN ISO 9001. BSI British Standards; 2008

[35] ABNT. NBR 15215-3: iluminação natural—Parte 3: Procedimento de cálculo Para a determinação da iluminação natural em ambientes internos. Rio de Janeiro: ABNT; 2005

[36] Hastings SR, Wall M. Sustainable Solar Housing: Strategies and Solutions. United Kingdom: Routledge; 2009

[37] USDE. Part I: Review of the Life-Cycle Energy Consumption of Incandescent, Compact Fluorescent, and LED Lamps. Life-Cycle Assessment of Energy and Environmental Impacts of LED Lighting Products. Washington (DC): U.S. Department of Energy; 2012

[38] IEA. Solar energy systems in architecture: Integration criteria and guidelines. Report T.41.A.2: IEA SHC Task 41. Solar Energy and Architecture. Solar Heating and Cooling Programme. Paris, France: International Energy Agency; 2012

[39] Raimo PA. Aquecimento de água no Setor Residencial [Dissertation]. São Paulo: Universidade de São Paulo; 2007

[40] Borges T. Síntese Otimizada de Sistemas de Aquecimento Solar de água [Dissertation]. Campinas: Universidade Estadual de Campinas; 2000

[41] Building integrated protovoltaics. Multifunctionality of BiPV. Available from: http://www.bipv.ch/index.php/en/products-en-top/multifunctionality-ofbipv [Accessed: August 22, 2018]

[42] Fthenakis V et al. Methodology guidelines on life-cycle assessment of photovoltaic electricity. Report IEA-PVPS T12-03:2011. International Energy Agency Photovoltaic Power Systems Programe; 2011

[43] Wild-Scholten MJ. Energy payback time and carbon footprint of commercial photovoltaic systems. Solar Energy Materials and Solar Cells. 2013;**296**:305-119. DOI: 10.1016/j.solmat.2013.08.037

[44] Ruther R. Edifícios Solares Fotovoltaicos: o Potencial da geração Solar Fotovoltaica Integrada a edificações Urbanas e Interligadas a Rede elétrica pública no Brasil. Florianópolis: LABSOLAR; 2004

[45] Pedrini A. Eficiência energética Em edificações e Equipamentos eletromecânicos. Rio de Janeiro: Programa Nacional de Conservação de Energia Eletrica; 2011

[46] Avaliação do mercado de eficiência energética no Brasil. Pesquisa de posse de equipamentos e hábitos de usos—ano base 2005—Classe Residencial—Relatório Brasil. Rio de Janeiro: Programa Nacional de Conservação de Energia Elétrica; 2007

[47] Meyers RJ, Williams ED, Matthews HS. Scoping the potential of monitoring and control technologies to reduce energy use in homes. Energy and Buildings. 2010;**563**:569-542. DOI: 10.1016/j.enbuild.2009.10.026

[48] Bartram L, Woodbury R. Smart homes or smart occupants? Reframing computational design models for the green home. In: Papers from the AAAI 2011 Spring Symposium—Artificial Intelligence and Sustainable Design. Palo Alto, California: Association for the Advancement of Artificial Intelligence; 2011

[49] Projeto Ekó House. Project Manual. São Paulo: Team Brasil; 2012

[50] Pallonetto F, De Rosaac M, Milano F, Finnac DP. Demand response algorithms for smart-grid ready residential buildings using machine learning models. Applied Energy. 2019;**239**(1):1265-1282. DOI: 10.1016/j.apenergy.2019.02.020

[51] Camilo HF, Udaeta MEM, Gimenes ALV, Grimoni JAB. Assessment of photovoltaic distributed generation—Issues of grid connected systems through the consumer side applied to a case study of Brazil. Renewable and Sustainable Energy Reviews. 2017;**71**:712-719. DOI: 10.1016/j.rser.2016.12.099

[52] Viana MS, Manassero Junior G, Udaeta MEM. Analysis of demand response and photovoltaic distributed generation as resources for power utility planning. Applied Energy. 2018;**217**:456-466. DOI: 10.1016/j.apenergy.2018.02.153

[53] Tschoppa D, Tianb Z, Berberichc M, Fand J, Perersd B, Furbo S. Large-scale solar thermal systems in leading countries: A review and comparative study of Denmark, China, Germany and Austria. Applied Energy. 2020;**270**: 114997. DOI: 10.1016/j.apenergy.2020.114997

[54] TTiempo. Madrid's temperature and relative humidity. 2021. Available from: https://pt.tutiempo.net/clima/2012/ws-82270.html [Accessed: July 17, 2021]

[55] IAG. Air temperature and relative humidity in São Paulo. 2021. Available from: http://www.estacao.iag.usp.br/Boletins/2012.pdf [Accessed: July 26, 2021]

[56] MME, EPE. Balanço Energético Nacional de 2020. Brasília: Ministério de Minas e Energia, Empresa de Pesquisa Energética; 2021

Chapter 7

Evaluation of Energy Efficiency of Buildings Based on LCA and LCC Assessment: Method, Computer Tool, and Case Studies

Suzana Domjan, Ciril Arkar, Rok Fink and Sašo Medved

Abstract

In this chapter, the development of a computer tool for the determination of nearly zero energy buildings (nZEB) metrics upgraded with life cycle assessment (LCA) and life cycle cost (LCC) indicators is presented, following the requirements of the Energy Performance of Buildings Directive (EPBD). The computer tool was developed for the assessment of new and renovated buildings to support the holistic decision-making process. The tool itself consists of two modules: the building description module (BDU), based on the national certification tool of buildings' energy performance, and the LCA tool (E^{tool}). BDU enables the assessment of energy needs, final energy demand, and primary energy needs. According to the EPBD, supporting standards was upgraded with the life cycle inventory database. The database includes data on predefined building materials, envelope components, heat generators, and energy carriers and is used by E^{tool} with which mid-point and end-point life cycle impact assessment can be done by taking into account impact groups and damage factors from IMPACT2002+ and ReCiPe methods. The LCC assessment module, which is also part of E^{tool}, was developed according to Commission Delegated Regulation No. 244/212. The use of computer tools is demonstrated through the case studies.

Keywords: energy performance of buildings directive, nearly zero energy buildings, lifecycle energy demand assessment, life cycle assessment, lifecycle cost assessment, computer tool

1. Introduction

The nearly zero-energy buildings (nZEB) requirements were introduced by the Energy Performance of Building Directive (recast) (EPBD recast) [1]. The EPBD supporting standards include the metrics and calculation procedures. The calculation starts with the determination of energy needs, followed by the determination of the required amounts of energy carriers produced on-site, nearby, or by distant systems.

The energy carriers should consist of a large share of renewable energies, which leads to lower primary energy needs. The directive explicitly requires that on the national level energy efficiency criteria must be set in a way to be cost-effective. Although the life cycle assessment (LCA) is voluntary, such approach can be very helpful in the decision-making process of building design. This challenging task should be performed in the early stage of the planning, requiring the relevant multidisciplinary knowledge [2]. A simplified computational tool can significantly help the implementation of EPBD in the planning process. Such a computer tool that enables building energy efficiency, the environmental impact of selected measures, and life cycle cost (LCC) assessment of new and renovated buildings was developed and is presented in this chapter.

2. Computer tool structure

Computer tool consists of life cycle energy efficiency (LCEA), environment impact (LCIA), and life cycle cost assessment (LCCA) routine, divided into two calculation modules, building description unit (BDU) and LCA tool (E^{tool}).

The BDU module enables the determination of:

- energy needs and final energy demand for operation of the building per energy carriers;
- environmental indicators in form of equivalents of pollutants of materials, building structures, building service systems, and energy carriers; and
- cost of materials, building structures and building service, and energy carriers.

The BDU is developed in a way that allows parallel analyses of two projects (for reference and designed buildings), allowing designers prompt and more convenient way for evaluation of proposed measures for increasing the energy efficiency of the designed building. For the same reason, designers could indicate separately which material, building structure, or building system will be included in the LCA. Such elements, marked as "LCA elements" are taken from the pre-designed database, but relevant data could be entered to form user-defined data.

After completing work in the BDU, data are automatically transferred into the second calculation module, the LCA evaluation module (E^{tool}) in which end-LCA results are determined and displayed taking into account additional user-selected LCA data, for example, lifetime and discount rate and compared for reference and current designed building. The main reason that the evaluation tool is divided into two modules is to use an existing highly distributed tool used for obligatory EPBD evaluation and certification of the buildings with more than 5000 users in Slovenia. A tool, called KI Energija [3] was co-developed by the authors of the presented chapter. Besides that, the tool was developed for use in high-school and master education courses through wizard-designed building service systems. The second module, E^{tool} was developed in an MS Excel environment because of built-in statistical functions and the ability to display results. The structure of the LCA computer tool is presented in **Figure 1**. The monthly method is used for the determination of energy needs for heating and cooling (ISO EN 52016-1 Energy performance of buildings – Energy needs for heating and cooling, internal temperatures, and sensible and latent heat loads – Part 1: Calculation

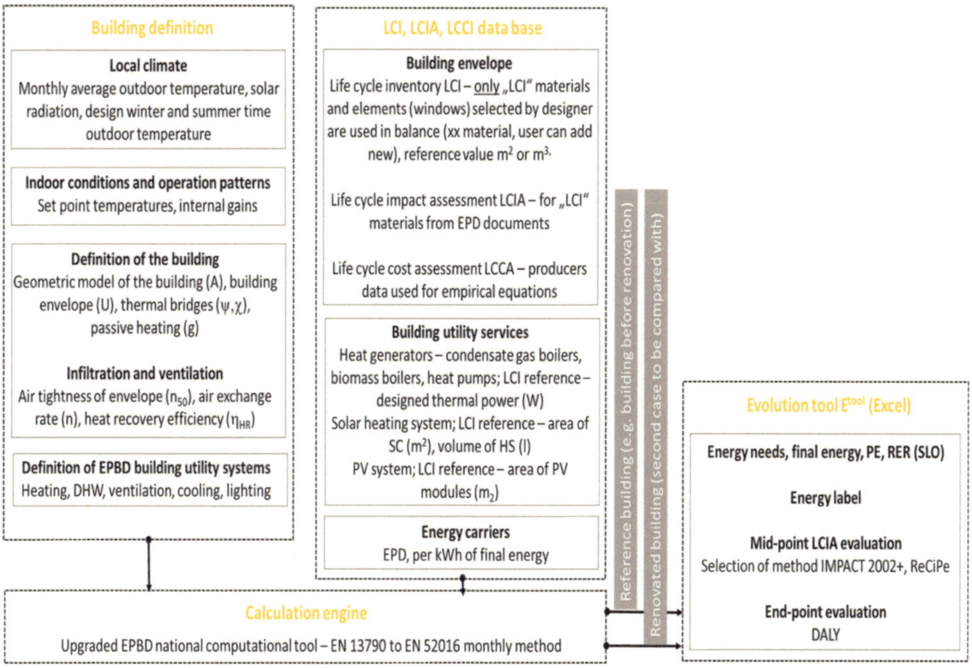

Figure 1.
Structure and interactions between BDU (upgraded EPBD national certification tool KI Energija) and E^{tool} (life cycle assessment tool).

procedure) and other EPBD supporting standards were used for final energy determination. Yearly methods for determining non-energy-related variables (e.g. emissions of pollutant equivalents) are used in the evaluation procedure.

2.1 National buildings' energy performance certification tool

The energy efficiency of the reference and designed building are determined by the following indicators:

- Energy needs for heating Q_{NH} and cooling Q_{NC}; monthly method according to EN ISO 52016-1 [4] (replacing EN ISO 13790 [5]) assuming constant set-point temperature for heating and cooling;

- Amount of each energy carrier as final energy demand for operation of installed building service for heating, cooling, domestic hot water (DHW) heating, ventilation, and lighting is determined per month and year. Several configurations of building service systems were pre-designed in the computer tool. An example of combined space heating and domestic water heating system is shown in **Figure 2**. In this way, the energy balance can be easier overviewed for each element of the system, and heat losses can be minimized most efficiently.

- Using primary energy factors and CO_2 emission factors, yearly primary energy needed for the operation of the buildings and CO_2 emissions are determined according to energy carrier demand. Because the tool is based on the current Slovenian national regulative, primary factors are not split into non- and

	Skupaj kWh/a
Toplota	
QNH,1	0
Qw	3264
Qf,h	7311
Qf,w	5442
Qove v Qf	8188
Qh,f,G1	7311
Qw,f,G1	5442
Qh,em,l	744
Qh,d,l	9617
Qw,d,l	2012
Qw,s,l,H1	454
Qh,g,l,G1	0
Qw,g,l,G1	0
Qrhh,em	0
Qh,d,rhh,aux	10
Qrhh,d	8813
Qrwh	1155
Qrww,d,aux	0
Qrhh,s,aux,G1	0
Qrww,s,aux,G1	0
Qrhh,g,G1	0
Qrww,g,aux,G1	0
Qrhh	8823
Električna energija	

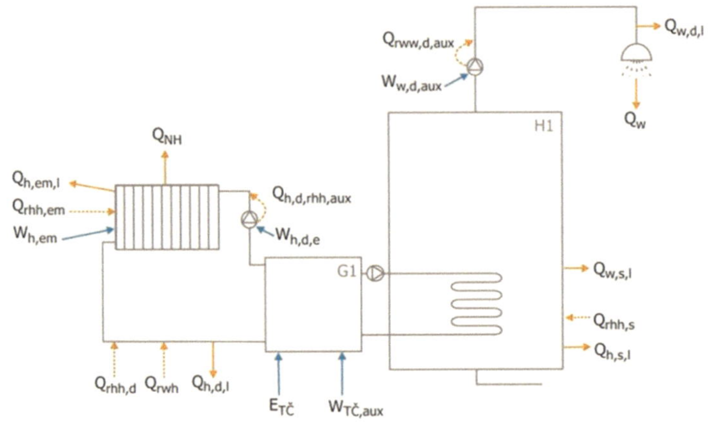

Figure 2.
An example of user's interface in national certification tool KI Energija with the presentation of energy balance for each element of heating and DHW heating system.

renewable energy ones, therefore, non-renewable primary factors for fuels and total primary energy factors for electricity are taken into account. This means that primary energy delivered by renewable energy sources (solar, environment heat) will be equal to zero.

3. Building description unit

The national EPBD certification computer tool was upgraded into the BDU by creating additional database files called "LCA" that includes information on the most recognized environmental impact indicators and cost. If the designer installs the "LCA" marked material into the building structure or the "LCA" marked component of the building service system, this will not only affect energy demand evaluation, but additional inventory data for LCIA and LCCA will be created. Because LCA indicators are developed per functional unit, the total value of each indicator (environmental impact or cost) is determined according to the building plan and stored in BDU. At the current stage of software development, inventory data are available for most commonly used construction materials, windows and doors as building structures, and heat generators as well as photovoltaic (PV) modules. Nevertheless, the inventory LCA database is open source and could be enlarged by the new elements with user-provided data (**Figure 3**).

3.1 Life cycle environmental impact assessment algorithm

The environmental impact indicators were chosen from Environment Product Declaration (EPD) certificates. Following damage categories are included: emissions of greenhouse gases causing global climate change weighted by Greenhouse Warming Potential (GWP) and expressed as CO_2 equivalent, emissions of gases that cause depletion of stratospheric ozone weighted by Ozone Depletion Potential (ODP) and

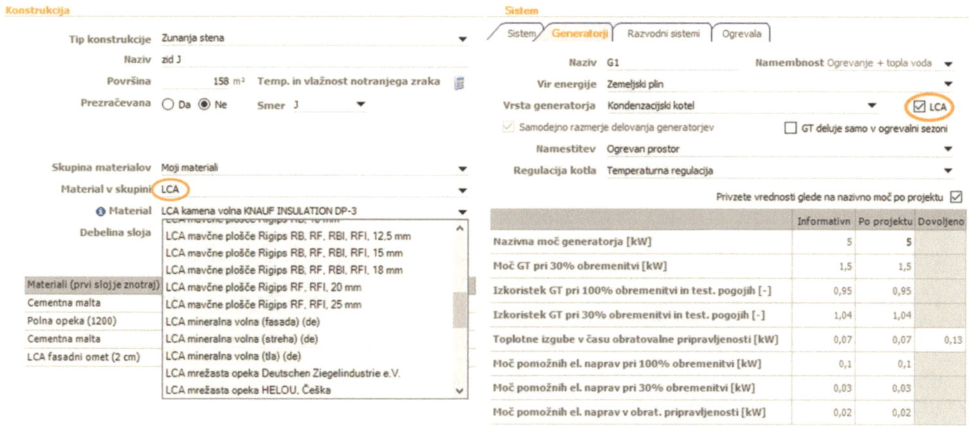

Figure 3.
If the designer wants to include certain materials (left), building structure, components of heat or electricity generation system (right), or particular energy carrier into the LCA assessment, it should be selected from the predefined "LCA" inventory database.

expressed as CFC-11 equivalent, emissions of gases that cause acidification of precipitation weighted by Acidification Potential (AP) expressed in SO_2 equivalent, eutrophication by weighted emissions by Eutrophication Potential EP as PO_4^{3-} equivalent, tropospheric ozone creation by Tropospheric Ozone Forming Potential (TOFP) as C_2H_4 equivalent, use of abiotic sources as Abiotic Depletion Potential – Elements (ADPE) as Sb equivalent and as Abiotic Depletion Potential – Fossil (ADPF) in MJ. Data from Ökobaudat [6], Environdec EPD Database [7], Eco-Platform [8], IBU [9], and manufactures data (i.e. Knauf Insulation [10]) were used in database integrated in BDU.

3.1.1 Life cycle environmental impact data of materials and building structures

Indicators presented in Chapter 3.1 are defined per reference unit. This is 1 m^3 of built-in material, except for thin layers, such as water vapor or wind barriers for which the reference unit is 1 m^2. BDU was adapted to calculate the total amount of built-in LCA materials and the total value of a particular environmental impact indicator. LCIA data of windows and doors are entered by default as such building structures are most commonly replaced as part of the energy renovation. To enable LCIA regardless of the size and type of the windows, regression models of each environmental impact indicator were developed taking into account the window glazing, spacer, and frame material. Factors are integrated into the BDU in the following form (as an example of greenhouse gas emissions):

$$GWP_w = A_w \cdot f_g \cdot GWP_g + \frac{A_w \cdot (1 - f_g)}{d_{frame}} \left(GWP_{frame} + GWP_{spac}\right) \left(kg\ CO_{2eq}\right) \quad (1)$$

where GWP_w is the impact factor of global climate change related to the window with area A_w (m^2), f_g is the ratio of glazing in the total window area, d_{frame} is the width of the frame (by default 0.1 m for wood and 0.15 m for plastic and metal frame), and GWP_g, GWP_{frame}, GWP_{spac} are specific impact factors per unit of glazing, frame, and spacer respectively. Default environment impact factors for windows and reference units are presented in **Table 1**.

	Unit	GWP	ODP	AP	EP	POCP	ADPE	ADPF
		kg eq CO_2	kg eq CFC11	kg eq SO_2	kg eq $(PO_4)^{3-}$	kg eq C_2H_4	kg Sb	MJ
Frame – wood	m	2.45	0.000000712	0.0176	0.00269	0.00874	0.0000360	74.33
Frame – Al	m	12.95	0.000005085	0.0554	0.00382	0.00339	0.0000994	146.90
Frame – PVC	m	8.07	0.000000986	0.0243	0.00267	0.00415	0.0001957	135.00
Window wing frame – wood	m	1.32	0.000000766	0.0195	0.00290	0.00925	0.0000403	80.94
Window wing frame – Al	m	12.44	0.000004885	0.0532	0.00367	0.00326	0.0000955	141.10
Window wing frame – PVC	m	9.09	0.000001202	0.0269	0.00302	0.00469	0.0002060	156.60
Glazing – double (0.75 W/m^2K)	m^2	37.52	0.000000534	0.1578	0.03022	0.01306	0.0002126	435.00
Glazing – triple (0.6 W/m^2K)	m^2	58.64	0.000000911	0.2446	0.04674	0.02026	0.0003254	695.60

Table 1.
Environment impact factors and reference units of window elements; data are average value gathered from oekodatbaudat.de and include A1–A3 LCA modules.

3.1.2 Life cycle environmental impact data of selected heat and electricity generators

Replacement of old heat generators and installing the solar thermal system or photovoltaic system are very common measures to increase energy efficiency and share renewable energy sources in buildings. LCIA data gathered from EPD databases [6–9] are integrated into BDU for the following on-site energy generators of buildings service systems: condensate gas boilers, biomass boilers, heat pumps, and solar thermal systems. Heat storage can also be included in LCIA and LCCA. For each LCIA indicator, the same form of regression model was developed and regression coefficients a_0, a_1, and a_2 were determined from the available database or research sources including A1–A3 LCA module. Coefficients were determined for boilers with design thermal power (as reference unit) between 20 and 400 kW, for heat pumps with design thermal power between 10 and 70 kW, and storage with the volume between 50 and 2500 liters. LCIA regression model was developed for PV modules with different PV cell technologies as well. In this case reference unit is the area of PV modules. Impact factors are integrated into the BDU in the following form (as an example of stratospheric ozone depletion potential):

$$
\begin{aligned}
ODP_{gen} &= a_{0,gen} + a_{1,gen} \cdot P_{gen} + a_{2,gen} \cdot P_{gen}^2 \quad (\text{kg CFC } 11_{eq}) \\
ODP_{sol} &= 1.25 \cdot a_{1,sol} \cdot A_{sc} \quad (\text{kg CFC } 11_{eq}) \\
ODP_{hs} &= a_{0,hs} + a_{1,hs} \cdot V_{hs} + a_{2,hs} \cdot V_{hs}^2 \quad (\text{kg CFC } 11_{eq}) \\
ODP_{pv} &= a_{1,pv} \cdot A_{pv} \quad (\text{kg CFC } 11_{eq})
\end{aligned} \tag{2}
$$

where $a_{0,x}$, $a_{1,x}$, and $a_{2,x}$ are regression coefficients for a particular building service system, P_{gen} (kW) is designed thermal power of heat generator, A_{sc} is the area of solar collectors (m^2), V_{hs} is the volume of heat storage (l), and A_{pv} is the area of PV modules (m^2). Default values of regression coefficients are shown in **Table 2**.

Heat generator – fuel oil					**Heat pump – water–water**				
kW		$a_{2,gen}$	$a_{1,gen}$	$a_{0,gen}$	**kW**		$a_{2,hp}$	$a_{1,hp}$	$a_{0,hp}$
GWP	kg eq CO_2	0.028	1.272	1566.316	GWP	kg eq CO_2	−0.911	258.13	−1088.2
ODP	kg eq CFC11	2.74E-13	1.24E-11	6.42E-08	ODP	kg eq CFC11	1.24E-07	−2.43E-06	2.11E-04
AP	kg eq SO_2	1.03E-04	4.67E-03	7.161	AP	kg eq SO_2	−0.002	0.676	−2.664
EP	kg eq $(PO_4)^{3-}$	9.27E-06	4.18E-04	0.601	EP	kg eq $(PO_4)^{3-}$	−2.72E-04	0.073	−0.305
POCP	kg eq C_2H_4	1.45E-05	6.64E-04	0.768	POCP	kg eq C_2H_4	−6.91E-04	0.175	−0.787
ADPE	kg Sb	2.33E-09	7.37E-08	0.054	ADPE	kg Sb	2.22E-05	−7.98E-05	5.81E-02
ADPF	MJ	0.319	14.443	18,151	ADPF	MJ	−18.73	5734.9	−25,366
Heat generator – kondens.					**Heat pump – ground heat exchanger**				
kW		$a_{2,gen}$	$a_{1,gen}$	$a_{0,gen}$	**kW**		$a_{2,hp}$	$a_{1,hp}$	$a_{0,hp}$
GWP	kg eq CO_2	0.0035	10.166	618.4	GWP	kg eq CO_2	0.093	550.585	116.167
ODP	kg eq CFC11	−1.00E-13	6.00E-10	6.00E-08	ODP	kg eq CFC11	2.46E-07	−7.36E-06	3.44E-04
AP	kg eq SO_2	5.00E-06	4.86E-02	3.6074	AP	kg eq SO_2	3.33E-04	1.164	0.729
EP	kg eq $(PO_4)^{3-}$	6.00E-07	4.00E-03	0.2837	EP	kg eq $(PO_4)^{3-}$	2.79E-05	1.20E-01	5.61E-02
POCP	kg eq C_2H_4	1.00E-06	4.10E-03	0.2605	POCP	kg eq C_2H_4	2.77E-05	2.69E-01	4.22E-02
ADPE	kg Sb	9.00E-07	1.60E-03	0.0835	ADPE	kg Sb	5.43E-05	−1.47E-03	9.82E-02
ADPF	MJ	4.33E-02	1.25E+02	7548.7	ADPF	MJ	1.006	19056.81	−2688.4
Heat generator – fuel oil					**Heat pump – water–water**				
kW		$a_{2,gen}$	$a_{1,gen}$	$a_{0,gen}$	**kW**		$a_{2,hp}$	$a_{1,hp}$	$a_{0,hp}$
GWP	kg eq CO_2	0.0008	17.158	626.8	GWP	kg eq CO_2	1.10E-01	39.655	213.8
ODP	kg eq CFC11	4.00E−14	9.00E-10	6.00E-08	ODP	kg eq CFC11	2.45E-07	7.35E-06	3.44E-04
AP	kg eq SO_2	4.00E-06	7.61E-02	3.8412	AP	kg eq SO_2	3.81E-04	7.97E-02	1.043
EP	kg eq $(PO_4)^{3-}$	3.00E-07	6.40E-03	2.991	EP	kg eq $(PO_4)^{3-}$	3.27E-05	8.31E-03	8.56E-02
POCP	kg eq C_2H_4	3.00E-07	7.00E-03	0.2658	POCP	kg eq C_2H_4	3.35E-05	1.94E-02	8.17E-02

Heat generator – fuel oil						**Heat pump – water–water**				
kW		$a_{2,gen}$	$a_{1,gen}$	$a_{0,gen}$		**kW**		$a_{2,hp}$	$a_{1,hp}$	$a_{0,hp}$
ADPE	kg Sb	9.00E-08	2.90E-03	0.0791		ADPE	kg Sb	6.45E-05	-1.92E-03	1.22E-01
ADPF	MJ	1.09E-02	2.11E+02	7649.3		ADPF	MJ	1.374	1441.48	−1466.2
Heat storage						**Heat pump – air-water**				
l		$a_{2,hs}$	$a_{1,hs}$	$a_{0,hs}$		**kW**		$a_{2,hp}$	$a_{1,hp}$	$a_{0,hp}$
GWP	kg eq CO_2	−1.00E-04	0.769	66.197		GWP	kg eq CO_2	−5.68E-14	45	0.1
ODP	kg eq CFC11	−7.00E-15	4.00E-11	3.00E-09		ODP	kg eq CFC11	1.19E-09	6.48E-05	1.67E-08
AP	kg eq SO_2	−7.00E-07	4.10E-03	3.54E-01		AP	kg eq SO_2	−3.57E-05	1.60E-01	−4.50E-03
EP	kg eq $(PO_4)^{3-}$	−4.00E+08	3.00E-04	2.21E-02		EP	kg eq $(PO_4)^{3-}$	3.57E-06	4.07E-02	3.50E-04
POCP	kg eq C_2H_4	−5.00E-08	3.00E-04	2.54E-02		POCP	kg eq C_2H_4	2.38E-06	1.55E-02	3.33E-04
ADPE	kg Sb	−2.00E-08	1.00E-04	0.0105		ADPE	kg Sb	2.38E-06	2.79E-02	1.33E-04
ADPF	MJ	−1.80E-03	10.215	879.7		ADPF	MJ	2.38E-02	590.929	3.333
Solar collector*		**flat**	**vacuum**		**PV**		**mono.**	**poli.**	**CdTe**	**CuInGaSe**
m^2		$a_{1,sol}$			**m2**		$a_{1,pv}$			
GWP	kg eq CO_2	107	112.70		GWP	kg eq CO_2	247.188	206.807	75.1	86.5
ODP	kg eq CFC11	3.22E-08	1.75E-08		ODP	kg eq CFC11	4.36E-05	4.18E-05	5.66E-06	6.28E-06
AP	kg eq SO_2	0.831	1.205		AP	kg eq SO_2	1.264	1.061	0.472	0.457
EP	kg eq $(PO_4)^{3-}$	0.036	0.046		EP	kg eq $(PO_4)^{3-}$	0.144	0.121	0.065	0.079
POCP	kg eq C_2H_4	0.043	0.061		POCP	kg eq C_2H_4	0.075	0.066	0.045	0.041
ADPE	kg Sb	0.019	0.051		ADPE	kg Sb	0.018	0.014	0.048	0.097
ADPF	MJ	1176	1263		ADPF	MJ	4156.635	3364.239	978.6	1079.4

**Note: Other components of the solar heating system weighting factor is taken into account [12].*

Table 2.
Environment impact factors and reference units of selected elements of building service systems; data represent the average value gathered from Ökobaudat [6], Environdec EPD database [7] and research publications [2, 11], and include A1–A3 LCA modules.

3.1.3 Life cycle environmental impact data of energy carriers

The database of environmental impact indicators of fuels is summarized from the EPD certificates gathered from Ökobaudat [6] and Environdec EPD database [7]. The reference unit is kWh of heat. The values of indicators consist of A1–A3 LCA modules. Impact indicators for electricity were determined based on EPD certificates of various technologies of electricity generation. Values are increased by a unified distribution factor. There is also an option to select electricity from the list of local electricity suppliers. Default values of impact factors of energy carriers are shown in **Table 3**. Besides default data, users can add their own data and this data will be available to use as an "LCA" marked element automatically.

3.2 Life cycle cost assessment

The important requirement of the recast EPBD is that EC Member States must set minimum requirements for energy performance of buildings in such a way that a cost-optimal solution is provided. The Directive [1] also defines the concept of a cost-optimal measure as a measure leading to the lowest total cost during the period of building operation. To assess the cost-effectiveness of energy efficiency measures, the LCCA module has been introduced in BDU. In the frame of the assessment, by discounting costs and savings, cash flow LCCn is determined in a pre-defined time period of n years and the investment with the highest positive cash flow can be found. In the BDU, the total cost of "LCA" elements is determined meanwhile cash flow is determined in E^{tool} in which the value of the investment is comparative, based on the

	Unit	GWP	ODP	AP	EP	POCP	ADPE	ADPF
		kg eq CO_2	kg eq CFC11	kg eq SO_2	kg eq $(PO_4)^{3-}$	kg eq C_2H_4	kg Sb	MJ
Natural gas	kWh	0.23890	1.73E-13	0.000177	0.0000267	0.0000297	1.12E-08	3.8740
Fuel oil	kWh	0.30700	3.17E-13	0.000338	0.0000434	0.0000422	1.15E-08	4.2300
Biomass (pellets)	kWh	0.00000	9.79E-13	0.000136	0.0000214	0.0000129	2.91E-09	0.2938
Electricity	kWh	0.27550	4.84E-07	0.002569	0.003032	0.0001734	7.03E-05	3.8998
District heating	kWh	0.26140	4.24E-13	0.000317	0.0000474	0.0000323	1.78E-08	3.0130
Sun	kWh	0	0	0	0	0	0	0
LG	kWh	0.26453	3.48E-13	0.000209	0.0000136	0.0000341	1.14E-08	4.1900
Environmental heat	kWh	0	0	0	0	0	0	0
Geothermal energy	kWh	0	0	0	0	0	0	0
Electricity (ECE, for households)	kWh	0.00616	2.06E-10	0.000023	0.0000350	0.0001301	1.70E-04	0.0454
Electricity (Gen-i)	kWh	0.46769	7.93E-07	0.004211	0.0049388	0.0002334	3.20E-05	6.6963

Table 3.
Environment impact factors and reference units of energy carriers.

calculated cost savings of the energy carriers in refurbished and reference building. The cash flow at the end of each year and the end of the calculation period n is determined by the equation:

$$LCC_n = \sum_{i=0}^{n} \left[(c_{inv} + c_{man})_i \cdot (1+d)^i + \sum_{j=1}^{m} c_{e,j} \cdot \frac{(1+e_j)^i}{(1+d)^i} \right] - V_{(n)}(€) \quad (3)$$

where i is the year numerator, n is LCA calculation period (years), and m is the number of energy carriers needed for the operation of the buildings. c_{inv} are investment costs (€), c_{man} yearly maintenance costs (/a), and $c_{e,j}$ is yearly costs of jth energy carriers (€/a), d is the discount rate, $c_{e,j}$ is forecasted yearly cost increase of jth energy carrier, and V(n) is the residual value of the built-in LCA element at the end of LCA calculation period. Guidelines accompanying Commission Delegated Regulation [13] suggested that for macroeconomic analysis, an annual discount factor of 3% should be assumed. The same document predicts the annual increase in energy carrier prices – 2.8% for natural gas and light heating oil prices, 2% for coal, and 9% increase in electricity prices (until 2030). The annual cost of maintenance of technical systems is assumed to be between 2 and 5% in cost-effectiveness studies [14]. The residual value of the measures is determined based on the expected lifetime of the measures. Standard EN 15459 [15] predicts the duration period of different energy efficiency measured – 50 years for thermal insulation on the building envelope, 30 years for building furniture, and 15 years for technical systems. According to the proposed LCA calculation period (30 years for residential building), this means that at the end of this period thermal insulation will have a residual discounted value of 16.5% of the investment value, taking into account the discount factor of 3%. The discounted residual value for building furniture will be EUR 0, while at least one replacement of technical systems will be required. For technical systems, the cost of replacement is discounted.

3.2.1 Cost database of materials and building structures

In parallel to inventory data of environmental indicators, costs are stored in BDU. For materials having reference units defined by the volume, costs in the database are defined as constant or as a linear function depending on the depth of the built material layer. Default costs are determined according to market research but could be modified by the user. The regression model for determination of costs of windows and doors was developed based on the hydraulic diameter. In the case of the window c_w, regression model is developed in form of:

$$c_w = b_{0,w} + b_{1,w} \cdot \frac{\overbrace{4 \cdot A_w}^{d_{w,H}}}{P_w} = b_{0,w} + b_{1,w} \cdot \frac{4 \cdot A_w}{\frac{(1-f_g) \cdot A_w}{d_{frame}} + 4 \cdot d_{frame}} \quad (4)$$

where $b_{0,w}$ and $b_{1,w}$ are regression coefficients, $d_{w,H}$ is the hydraulic diameter of window (m), A_w is window area (m^2), P_w in window perimeter (m), f_g is the ratio of glazing in the total window area, d_{frame} is the width of the frame (m). The regression model is valid for the windows with an area up to 4 m^2. In **Figure 4** costs of market available windows with wood frame and according to the hydraulic diameter of the

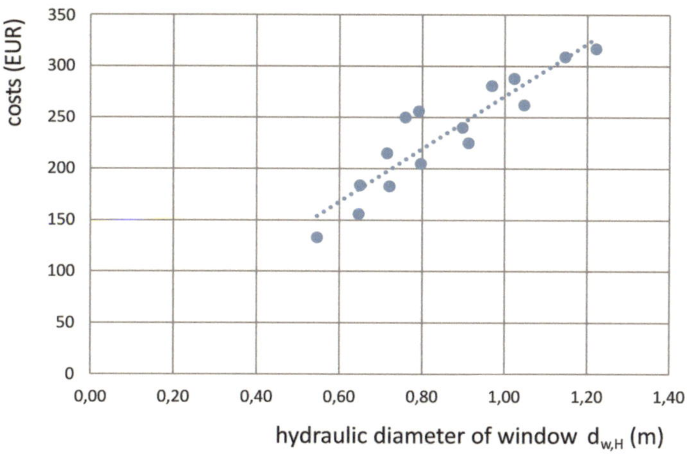

Figure 4.
Cost of windows with a wooden frame and two-layer glazing according to hydraulic diameter of the window; data gathered from Slovenian market overview [2].

window are shown. Data for double and triple glazing, as well as for wooden, plastic, and metal frames were gathered.

3.2.2 Cost database of building service systems elements

Investment cost of elements of building service systems integrated into BDU as default values are available for the following components: condensation gas boilers, biomass boilers, A/W, S/W and W/W heat pumps, solar heating systems with heat storage, mono, poly, CdTe, and CIGS PV modules. Regression models in similar form as environment impacts indicators (Eq. (2)) were developed with new regression coefficients as presented in **Table 4**. Regression coefficients were determined for boilers with design heating power 10–136 kW, for heat pumps thermal power from 5 to 90 kW, and for heat storage with volume 50–2500 liters and 1 m^2 of solar collector or PV module area.

3.2.3 Cost database for energy carriers

The cost of energy carriers was determined per kWh from data published by the Statistical Office of the Republic of Slovenia [16] and the Slovenian market price overview.

4. Life cycle assessment tool

BDU forms necessary data needed for LCEA, LCIA, and LCCA. It is designed in a way that two selected projects' data can be exported in the LCA evaluation tool E^{tool}, one as a reference and the other as a designed one. This allows immediate evaluation of proposed measures for increasing the energy efficiency of buildings. E^{tool} was developed in MS Excel software. Following the requirements of EPBD and content of environmental product declarations (EPD) the LCA metrics includes presentation of:

Heat generator – wood chips					**PV – mono.**				
		$b_{2,gen}$	$b_{1,gen}$	$b_{0,gen}$			$b_{2,pv}$	$b_{1,pv}$	$b_{0,pv}$
c_{gen}	EUR	−0.1836	93.984	3894.6	c_{pv}	EUR	0	260	0
Heat generator – kondens.					**PV – poli.**				
		$b_{2,gen}$	$b_{1,gen}$	$b_{0,gen}$			$b_{2,pv}$	$b_{1,pv}$	$b_{0,pv}$
c_{gen}	EUR	0.3224	−5.3907	1771.3	c_{pv}	EUR	0	230	0
Heat generator – fuel oil					**PV – CdTe**				
		$b_{2,gen}$	$b_{1,gen}$	$b_{0,gen}$			$b_{2,pv}$	$b_{1,pv}$	$b_{0,pv}$
c_{gen}	EUR	−0.1554	16.047	4209.4	c_{pv}	EUR	0	200	0
Heat pump – geosonde					**PV – CuInGaSe**				
		$b_{2,hp}$	$b_{1,hp}$	$b_{0,hp}$			$b_{2,pv}$	$b_{1,pv}$	$b_{0,pv}$
c_{hp}	EUR	−1.2566	947.84	1742	c_{pv}	EUR	0	200	0
Heat pump – ground heat exchanger					**Solar collector – flat**				
		$b_{2,hp}$	$b_{1,hp}$	$b_{0,hp}$			$b_{2,sol}$	$b_{1,sol}$	$b_{0,sol}$
c_{hp}	EUR	−1.2566	447.84	1742	c_{sol}	EUR	0	312.5	0
Heat pump – water–water					**Solar collector – vacuum**				
		$b_{2,hp}$	$b_{1,hp}$	$b_{0,hp}$			$b_{2,sol}$	$b_{1,sol}$	$b_{0,sol}$
c_{hp}	EUR	−1.2566	247.84	1742	c_{sol}	EUR	0	520	0
Heat pump – air-water					**Heat storage**				
		$b_{2,hp}$	$b_{1,hp}$	$b_{0,hp}$			$b_{2,hs}$	$b_{1,hs}$	$b_{0,hs}$
c_{hp}	EUR	−0.3418	282.63	3377.4	c_{hs}	EUR	5.00E-05	0.4354	453.98

Note: Assembly costs are taken into account as 20% (for heat pumps and solar collectors) and 10% (for the rest of the heat generators, PV, and heat storage) of investment costs.

Table 4.
Regression coefficients in regression cost models of selected elements of building service systems.

- LCEA – yearly specific energy needs for heating (Q'_{NH}), final energy (Q'_f) for the operation of EPBD building service systems, primary energy needed (Q'_p), and renewable energy ratio in delivered (final) energy are shown as the specific value per unit of useful area (**Figure 5**). These values allow the designer to overview the fulfillment of nZEB requirements. On the second level of LCEA (**Figure 6**), the absolute energy demand is shown, and delivered (final) energy is presented by energy carriers. Besides energy demand, emission of CO_2, as well as the emission of greenhouse gasses expressed as CO_2 equivalent is shown as the most recognizable environment impact indicators. Meanwhile, emissions of CO_2 are determined by energy carrier use, CO_{2eq} includes LCA emissions (A1–A3) resulting from the implementation of measures taken to increase the energy performance of the building. The impacts of all building elements taken from the "LCA" database or marked as "LCA" are summarized. For analyzed (e.g. renovated) buildings, the decrease of energy demand can be compared with the embodied energy of "LCA" elements through user-selected calculation period. Data of embodied energy is taken as the value of Abiotic Depletion Potential – Fossil (ADPF) environmental impact indicator of "LCA" elements from modules A1–A3.

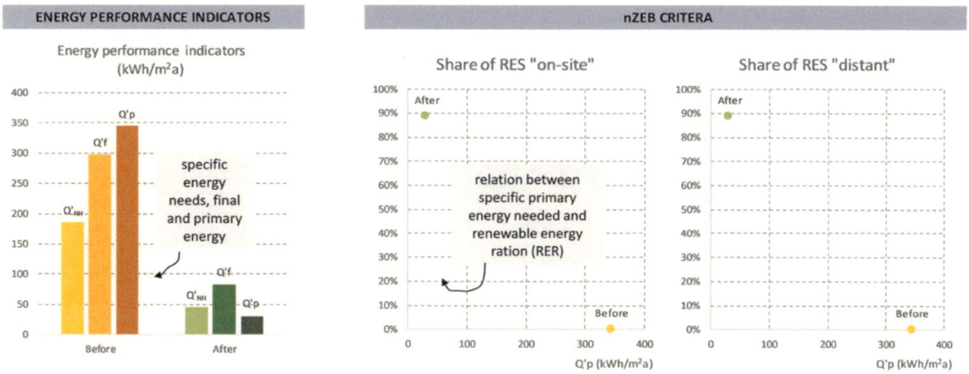

Figure 5.
Display of nZEB energy efficiency metrics at the base level of LCEA.

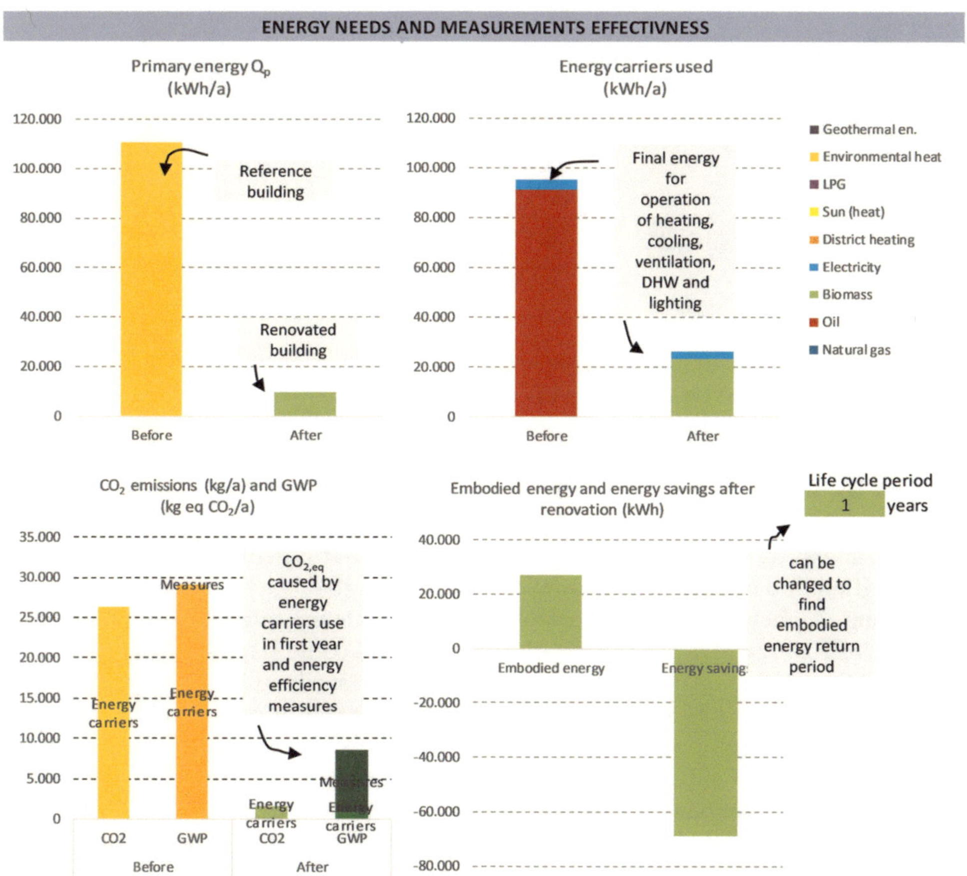

Figure 6.
Second level of LCEA metrics displayed in E^{tool}.

- LCIA – LCA environmental impact assessment is performed in three phases – through classification, characterization, and normalization phase (**Figure 7**). In the classification phase material and energy flows as well as the emission of

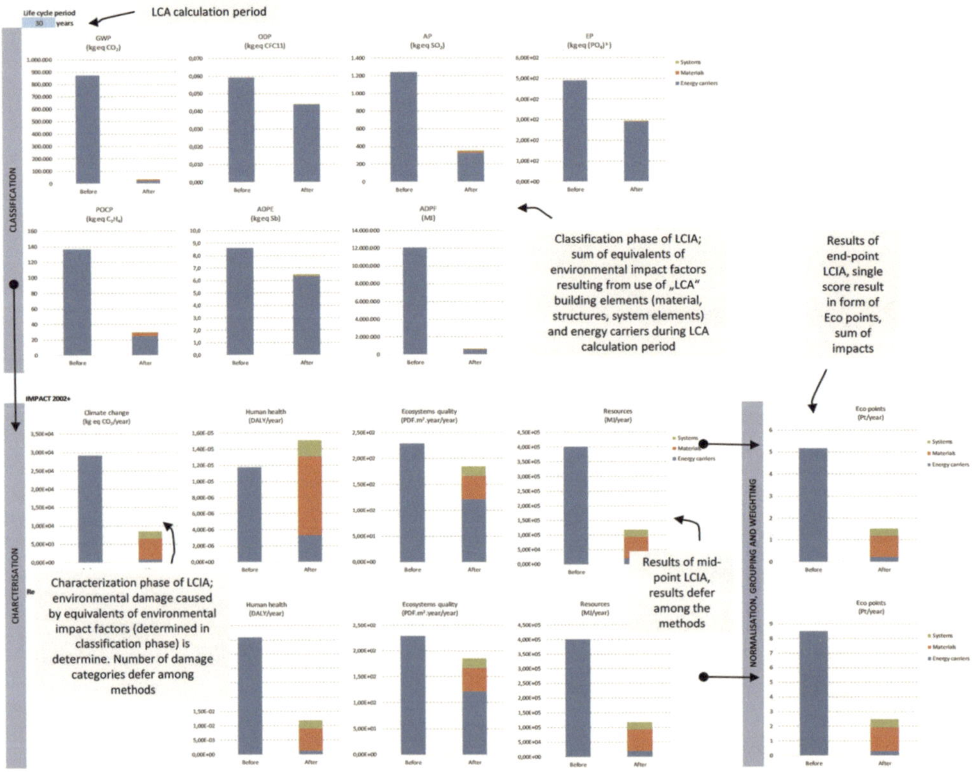

Figure 7.
Results of LCIA metrics.

pollutants equivalents related to LCA building elements, including energy carriers, are summarized during the user-selected LCA calculation period into seven pre-selected impact categories. Values are presented as physical quantities (e.g. kg, MJ). In the characterization phase sum of environmental impacts expressed by equivalents (e.g. AP or EP) are weighed by impact factors (e.g. global warming potential GWP_{100} of particular greenhouse gas) and classified into damage categories. The number of damage categories defers among the methods. As most commonly used, damage categories included in IMPACT 2002+ [17, 18] and ReCiPe [19] method could be evaluated in E^{tool}. IMPACT 2002+ consists of four damage categories: climate change (global warming), human health measured in DALY (Disability Adjusted Life Years), ecosystem quality measured as potential loss of ecosystems as a consequence of acidification and eutrophication and expressed as PDF (Potentially Disappeared Fraction), and damage to reserves of natural resources expressed in MJ. ReCiPe method assessed environmental impact only through three damage categories because global warming is included through the human health damage category. At this point, results are presented as mid-point environmental impacts to the global environment (e.g. DALY per year or MJ per year). By normalization, impacts of the analyzed building (reference and designed) are compared to the total environmental impacts in the reference system e.g. European Union and total impacts could be normalized to each person, with an assumed number of inhabitants 410×10^6. Mid-point LCIA results in E^{tool} are presented as total

Figure 8.
Results of LCCA metrics.

impacts. By normalization, the end-point in form of a single score can be evaluated by Eco Point per year (PT/y). All indicators are presented in a way that shows the contribution of each group of LCA elements (energy carrier, materials, and building service systems).

- LCCA – It is displayed by values presented in **Figure 8**; taking into account user-defined discount factor, the discounted price of energy carrier, and yearly cost of maintenance of "LCA" elements (as % of investment cost), cash flow diagram is constructed. A simple and discounted payback period is presented as well. The cash flow diagram is shown for the period of 30 years, the period defined for LCA of buildings in [13]; total LCCs for user-selected calculation period are shown next. Costs are shown for each LCA element (energy carriers, investment in materials for renovation, and investments for more efficient building system components). Only building service components presented LCIA can be analyzed. Cost-effective measures can be found from this data. Next, specific (per m^2 of useful area) lifetime costs of investments, building operation, and maintenance related to specific primary energy demand are shown, enabling the designer to choose the most cost-effective measures as requested in recast EPBD. The last LCCA result presents the cost of $eqCO_2$ emissions in the lifetime period as a macroeconomic indicator [11]. Because the cost of a unit of $eqCO_2$ emissions is dynamic, the first year of the calculation period must be defined by the user.

5. Case studies

5.1 Life cycle assessment of energy retrofitting of a public building

Public buildings must fulfill stricter measures and therefore energy retrofitting should be done even more carefully to justify proposed solutions beyond costs. According to that the implementation of LCIA into the decision-making process will be crucial for fulfilling the climate mitigation targets.

As an example, the assessment of energy retrofitting measures of the hospital is presented. District heat is used for heating and preparation of hot water. Hospital has a useful area A_u 7405 m^2 and energy needs for heating Q'_{NH} 1614 kWh/(m^2a). At current (reference) conditions, the primary energy needed for the operation of the building service systems Q_p is 1.865,362 kWh/a. Based on the parametric analysis, the planner decides on the proposed measures, and after the choice of measures, the energy, environmental, and cost assessment of the measures is carried out (**Figure 9**).

Following measures were chosen: windows replacement (U_w 3 W/m^2K → 1.1 W/m^2K), thermal insulation of the facade (U_{wall} 1.3 W/m^2K → 0.168 W/m^2K), and thermal insulation of the ceiling to the unheated attic (U_{roof} 0.957 W/m^2 → 0.094 W/m^2K). The mechanical ventilation with heat recovery was not included.

After the energy retrofitting, the BDU shows the following results, the specific energy needs for heating will be reduced by 75% (Q'_{NH} 161 kWh/m^2a → 39 kWh/m^2a), the final energy by 68% (Q'_f 220 kWh/m^2a → 70 kWh/m^2a), and the required specific primary energy for the operation of the building by 58% (Q'_p 252 kWh/m^2a → 107 kWh/m^2a) (**Figure 10**, left). The use of district heating heat and DHW will be reduced from 1380 to 320 MWh/a (**Figure 10**, middle). At this point, reference and

Figure 9.
Hospital building in Ljubljana.

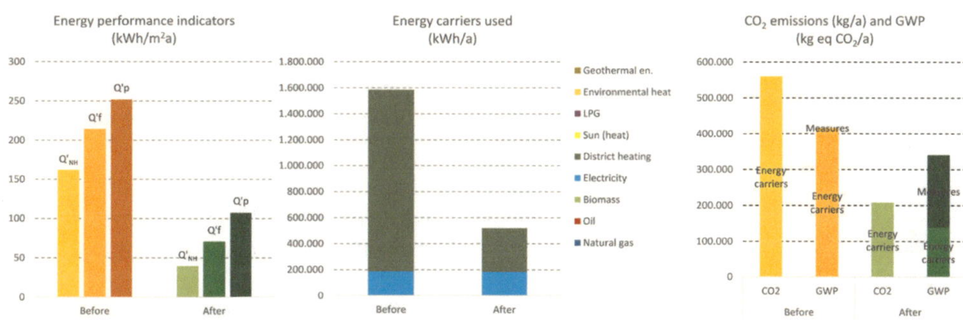

Figure 10.
Results of energy efficiency analysis: Energy performance indicators (left), energy carriers (middle), and CO_2 emissions and GWP (right) for reference project (before) and retrofitted building (after).

renovated building data were exported to E^{tool} for environmental and cost assessment. The results are presented for reference (before) and retrofitted building (after). The CO_2 emissions will decrease by 345 tons per year. The greenhouse gases (GHG) emissions caused by measures will be 202 tons of eqCO_2 per year, nevertheless, in the following years, the GHG emissions will be lower by 278 tons of eqCO_2 each year. Example shows that even measures can be justified according to the environmental impact, as total GHG emissions will be lower compared to the current state (**Figure 10**, right).

The comparison of embodied energy and energy savings of energy efficiency measures shows that "energy payback time" will be shorter than 1 year, which indicates that proposed materials and technologies are sustainable (**Figure 11**).

LCIA analysis shows that all environmental indicators are significantly improved during the assessment period (selected duration of 30 years). It can be seen that the

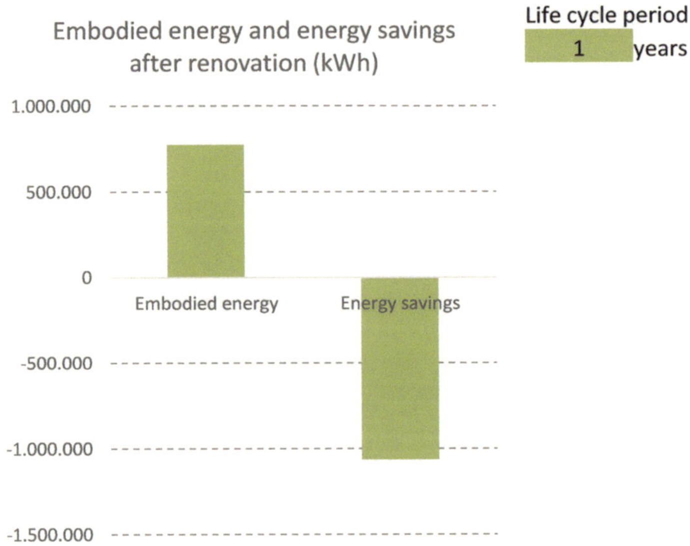

Figure 11.
Comparison of embodied energy in materials and technologies proposed for energy retrofitting and energy savings after 1 year of building operation.

impact of energy efficiency measures (presented as materials) on total GWP emissions is less than 5%, while the impact of measures on the environment is the largest for ODP, while it is smallest for ADP (**Figure 12**).

The reference building causes the population of the EU (431×10^6 inhabitants) 0.58 years of less quality living (DALY, ReCiPe), while the retrofitted building will cause 0.48 DALY in the first year, and 0.19 DALY/a in the following years (**Figure 13**, left). The number of Eco points after retrofitting resulting from the use of energy carriers will be reduced from 68 to 59 Pt/a in the first year and to 24 Pt/a in the rest of the calculation period (**Figure 13**, right).

The LCC results are shown in **Figure 14**. Taking into account the user-defined discount factor d 3% and energy price factor e 2.8%, and assumed maintenance costs of 0.5% of the investment per year, the payback period of the proposed measures will be 16 years, while the cost of energy carriers will be reduced from the current 391 to 268 €/m^2 of the useful building area (**Figure 14**).

5.2 Comparison analysis of on-site heat and electricity generators in a single-family building

The study case illustrates the process of evaluation technologies for heating and domestic water heating (DHW) as well as electricity generation in a single-family building with a useful area of 92 m^2. The buildings are designed according to the passive buildings criteria. The building is mechanically ventilated with a heat recovery system with an efficiency of 75%; the specific power of the fans $P_{v,dov}$ and P_v are 0.31 W/(m^3/h). The energy needs for heating Q'_{NH} are 11.9 kWh/(m^{2}a). The specific power of the built-in lamps is 3 W/m^2. In the reference building, the biomass pellets boiler is installed and connected to heat storage of floor heating system (600 l) and heat storage for DHW (300 l).

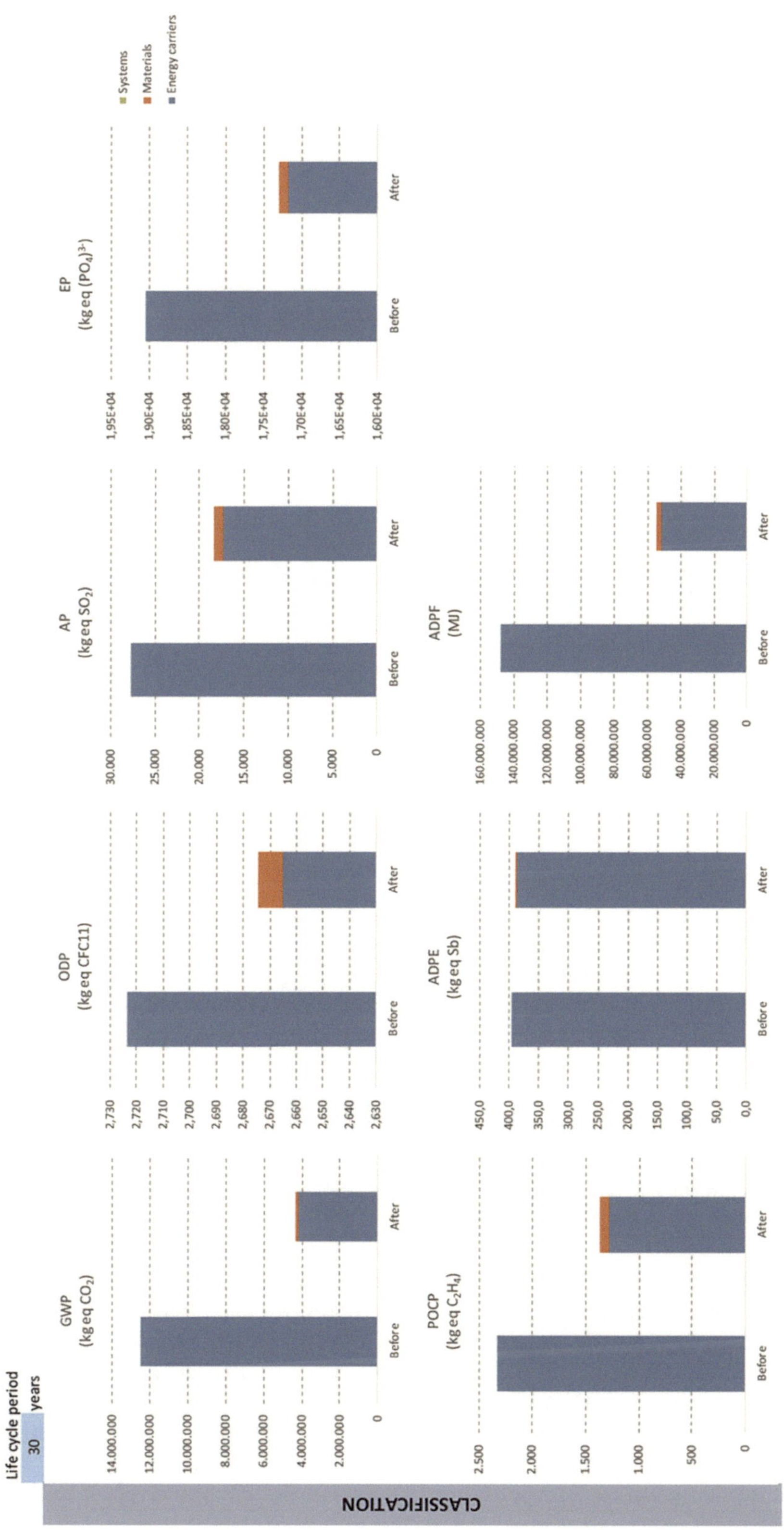

Figure 12.
LCIA analysis results of energy efficiency measures after the classification phase.

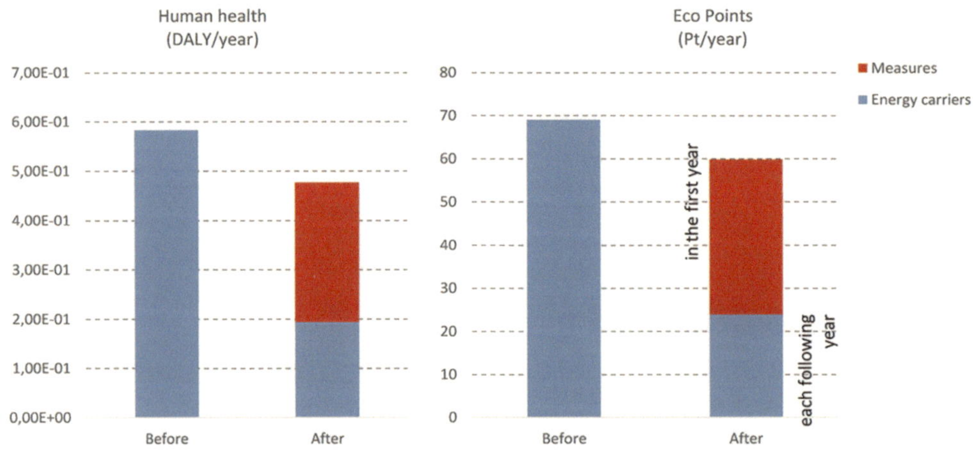

Figure 13.
LCIA analysis results of energy efficiency measures after the characterization phase.

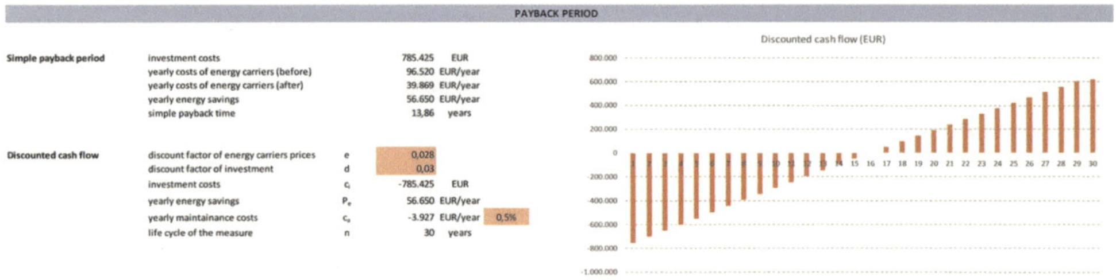

Figure 14.
Payback period analysis of the investment.

The following alternative technologies were analyzed:

- Case 1: heat pump (air-water) with rated heat power 5 kW, heat storage in the DHW system with a capacity of 300 liters;
- Case 2: natural gas condensing boiler and solar collectors, the surface of vacuum solar collectors is 7.5 m^2; the volume of the heat storage is 300 liters;
- Case 3: natural gas condensing boiler, with flow-through DHW heating, and a PV power plant with a power of 1.75 W_p, connected to the grid;

Energy efficiency analysis (LCEA). While the specific energy needs for heating Q'_{NH} are the same for all cases, the specific final energy demand for the operation of building service systems Q'_f is the smallest for Case 3 (42 kWh/(m^2a)), and approximately the same for Cases 1 and 2 (51.3 and 53.4 kWh/(m^2a)), and the highest in the case of a reference building (63 kWh/(m^2a)) (**Figure 15**).

The share of renewable energy sources (RES) for Cases 1 and 2 is provided from solar energy and environmental heat, while the required share of RES in Case 3 is provided by the transmission of electricity produced from the PV power plant to the grid (**Figure 16**).

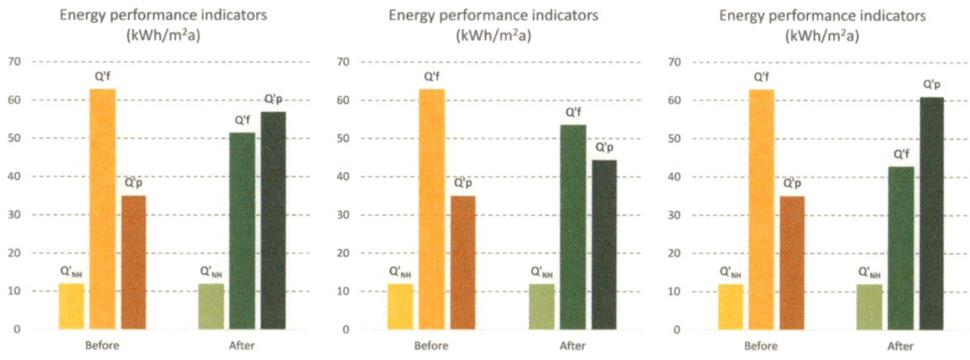

Figure 15.
Energy efficiency indicators for Case 1 (left), Case 2 (middle), and Case 3 (right) in comparison with the reference case (before).

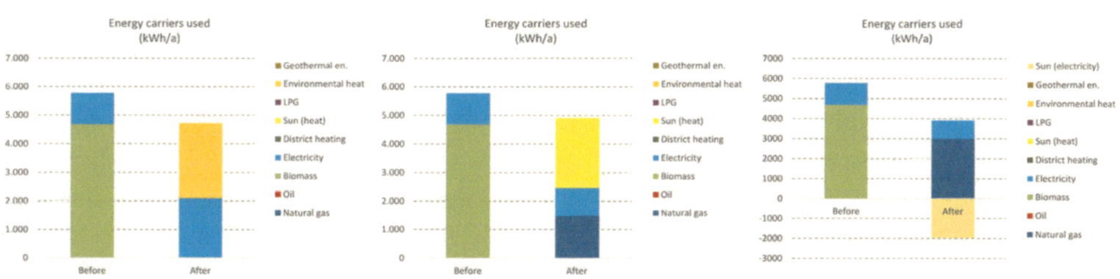

Figure 16.
Structure of energy carriers for Case 1 (left), Case 2 (middle), and Case 3 (right) in comparison with the reference case.

The largest difference in the embodied energy relative to the reference case is in Case 3, and the smallest in Case 1. In Case 3, the difference in total delivered energy is the largest also for the 30-years period (**Figure 17**).

Annual CO_2 emissions according to the Slovenian national legislation [20], are approximate two times higher as in the reference case, and the lowest in Case 2 (750 kg/a), i.e. by 35% compared to Case 1 and by 25% compared to Case 3. The classification of technologies according to GHG emissions (GWP) is the opposite, due to the lower use of energy carriers and the high share of RES in the electricity mix in last years in Slovenia (**Figure 18**). If another electricity supplier was selected, the GWP emissions of Case 1 would be close to 0.

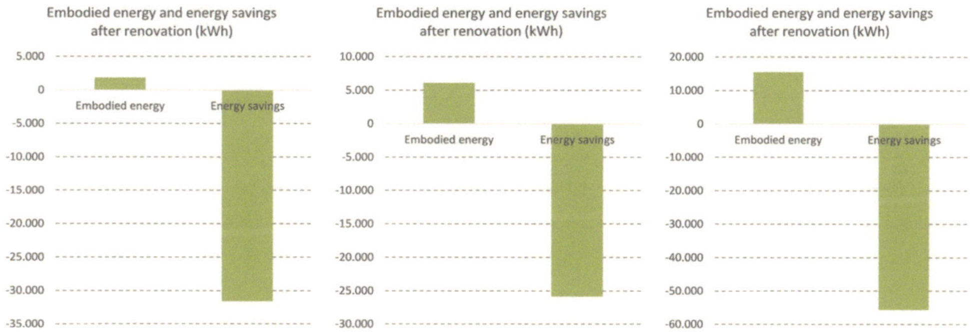

Figure 17.
Embodied energy for Case 1 (left), Case 2 (middle), and Case 3 (right).

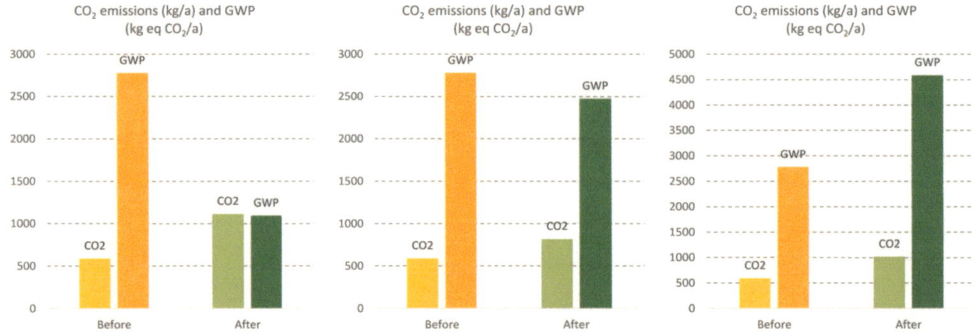

Figure 18.
Annual CO_2 and GWP emissions for Case 1 (left), Case 2 (middle), and Case 3 (right) in comparison with the reference case (before).

Environmental impact analysis (LCIA). The comparison of environmental impact was done based on Eco points of heat generators and delivered energy after the first year of operation. Compared to the reference case (pellet biomass boiler) with an impact of 0.462 Pt, the gas boiler with solar thermal collectors (Case 2) has approximately the same impact (0.427 Pt). The impact is approximately half of that in the case of the heat pump (Case 1, 0.199 Pt) and doubled in the case of the gas boiler with PV (Case 3, 0.870 Pt) (**Figure 19**). The difference mainly results from the environmental pressures caused by the use of materials and the production of system elements. After the first year of operation, damage to the environment is caused only by the use of energy carriers. The use of biomass causes the lowest yearly environmental impact (0.083 Pt/a). The impact of Cases 1 and 2 is higher for 25% and the impact of Case 3 is almost doubled (0,186 Pt/a). This analysis confirmed that LCIA is a very meaningful approach when choosing technologies for nZEB.

Cost analysis (LCCA). Assuming 30 years of operation, the total costs (investments and energy carriers) of reference case, Cases 2 and 3 are more or less the same (between € 18,300 and € 18,500), while total costs of Case 1 are lower by 17% (€ 15,300). The cost of energy carriers is close to the investment for Cases 1 and 3, whereas the investment represents 2/3 of the total costs over the 30-years period for Case 2 (**Figure 20**). The cost input parameters are presented in Section 3 (LCCA).

Macroeconomic costs, evaluated on the basis of eqCO_2 emissions costs [13] over the 30 years of operation, are the lowest at the reference case 610 €, for Case 1675 €, for Case 2750 €, and for Case 31,150 € (**Figure 21**). These ratios would be reasonable to use for creating public non-refundable financial incentives.

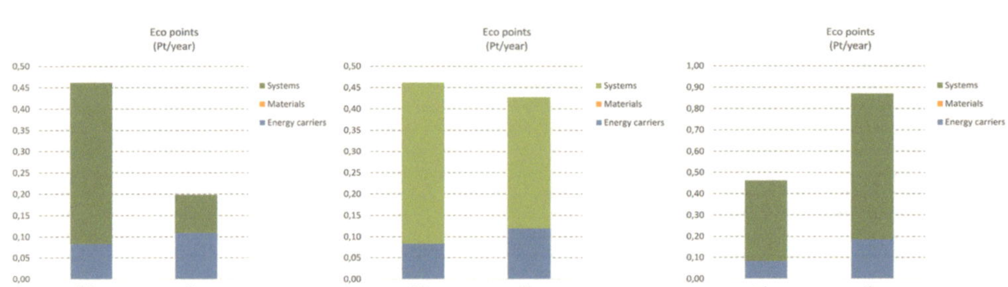

Figure 19.
Eco points for the first year of operation for Case 1 (left), Case 2 (middle), and Case 3 (right) in comparison with the reference case (before).

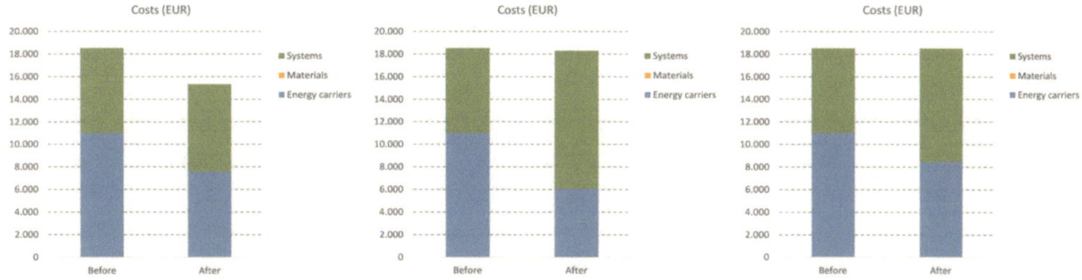

Figure 20.
Costs of the investment and 30-years operation for Case 1 (left), Case 2 (middle), and Case 3 (right) in comparison with the reference case (before).

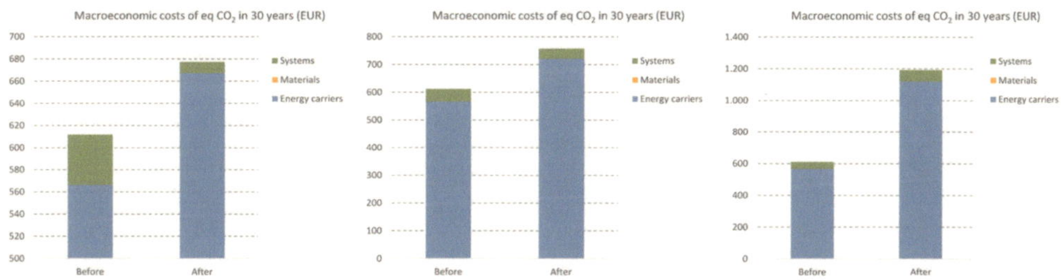

Figure 21.
$EqCO_2$ emissions costs for the 30-years operation period for Case 1 (left), Case 2 (middle), and Case 3 (right) in comparison with the reference case (before).

5.3 Optimization of multi-family building's energy retrofitting

Retrofitting of multifamily buildings in Ljubljana includes improvement of building's envelope. Building with a useful area 1950 m^2 (**Figure 22**) is heated by a district heating system. The existing building with the brick wall without thermal insulation (U_{wall} = 0.986 W/m^2K) and double paned glass windows with wooden frames

Figure 22.
Multifamily building in Ljubljana analyzed in the case study.

(U_w = 3.0 W/m^{2}K) was taken as a reference project. The ceiling toward the unheated attic was already insulated. The energy needs for heating Q'_{NH} of the reference project is 147.7 kWh/m^{2}a.

Optimization was made according to the specific costs of investment (including all façade layers and labor costs) and energy carriers over the 30-years period. It was found that maximum cost savings of 52 € per m^2 of the useful area can be achieved with thermal insulation's thickness of 25 cm. Such measure will result in decreasing primary energy demand from 251.6 to 196.8 kWh per m^2 of useful area, taking into account all installed building service systems. The cost of eqCO_2 emissions decreases up to the much larger thickness of thermal insulation and no optimum value can be determined. This means that according to the macro-economic cost of eqCO_2 emissions there is no need to limit subsidies based on thermal insulation thickness or U-value of building structures (**Figure 23**).

Figure 23.
Thermal insulation thickness optimization based on the criteria of cost-effectiveness in the life-cycle.

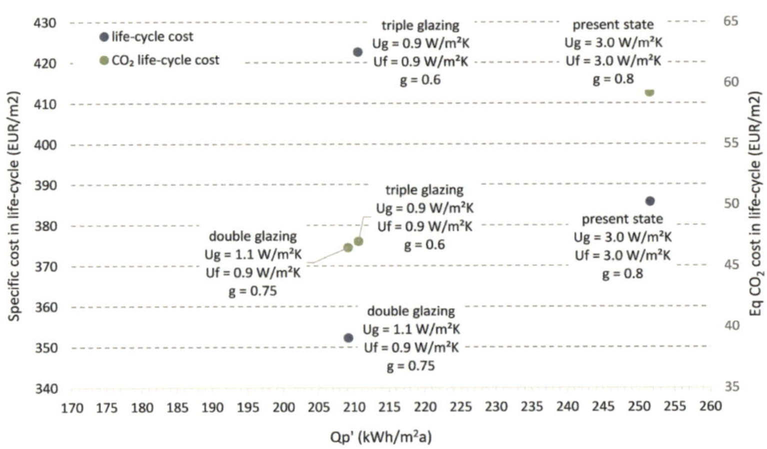

Figure 24.
Windows replacement optimization based on decreasing primary energy demand and the criteria of cost-effectiveness in the life cycle.

The impact of the replacement of windows was evaluated based on heat losses and solar gains. The life-cycle cost assessment showed that double glazed windows provide savings of 33 € per m^2 of useful area, while triple glazed windows are not cost-efficient and besides that such windows do not decrease primary energy demand more than double glazed windows (**Figure 24**). The macroeconomic costs of $eqCO_2$ emissions also give priority to double-glazed windows.

6. Conclusions

Through the use of the developed computer tool and showed cases, it can be pointed out that life-cycle assessment significantly helps in the decision-making process. Design and evaluation of nZEB metric that is nowadays focused on energy balance should be broadened to other aspects of assessment, including environmental impact and cost assessment, all based on life-cycle approach. It can be seen that different approaches give different optimal solutions. Therefore, the designer would not be the one to decide about the aspect of optimization, the weighting factors or, as we propose, the optimal solutions should be found according to the lowest macroeconomic costs of CO_2 emissions. In this way, subsidies schemes can be defined and contributions to the decarburization of the building sector will be presented most efficiently.

Author details

Suzana Domjan[1*], Ciril Arkar[1], Rok Fink[2] and Sašo Medved[1]

1 Faculty of Mechanical Engineering, Laboratory for Sustainable Technologies in Buildings, University of Ljubljana, Ljubljana, Slovenia

2 Faculty of Health Sciences, University of Ljubljana, Ljubljana, Slovenia

*Address all correspondence to: suzana.domjan@fs.uni-lj.si

References

[1] Directive 2010/31/EU of the European Parliament and of the Council of 19 May 2010 on the energy performance of buildings (recast). Luxembourg: Official Journal of the European Union. **L 153/13**:2010

[2] Domjan S. Parametrični modeli za večkriterijsko analizo skoraj nič energijskih stavb (Parametric models for multi-criteria analysis of nearly zero energy buildings) [master thesis]. Ljubljana: Faculty of Mechanical Engineering, University of Ljubljana; 2016

[3] Medved S et al. KI Energija 2014 (v 4.2.6.0). Škofja Loka: Knauf Insulation; 2014

[4] EN ISO 52016-1:2017 Energy performance of buildings—Energy needs for heating and cooling, internal temperatures and sensible and latent heat loads—Part 1: Calculation procedures

[5] EN ISO 13790:2008 Energy performance of buildings—Calculation of energy use for space heating and cooling

[6] Bundesministerium des Innern, für Bau und Heimat (BMI). Ökobaudat [Internet]. Available from: https://www.oekobaudat.de/ [Accessed: February 15, 2019]

[7] EPD International AB. Environdec EPD Database [Internet]. Available from: https://www.environdec.com/EPD-Search/ [Accessed: February 2, 2019]

[8] ECO Platform AISBL c/o Construction Products Europe AISBL. Eco-Platform [Internet]. Available from: https://www.eco-platform.org/ [Accessed: February 2, 2019]

[9] Das Institut Bauen und Umwelt. IBU EPD. [Internet]. Available from: https://ibu-epd.com/ [Accessed: February 2, 2019]

[10] Knauf Insulation. Knauf Insulation EPDs [Internet]. Available from: https://www.knaufinsulation.com/downloads/environmental-product-declarations-epd [Accessed: February 2, 2019]

[11] Pahor M. Parametrični model rabe toplote v stavbah (Parametrical models of energy needs of buildings) [diploma work]. Ljubljana: Faculty of Mechanical Engineering, University of Ljubljana; 2013

[12] Koroneos CJ, Nanaki EA. Life cycle environmental impact assessment of a solar water heater. Journal of Cleaner Production. 2012;**37**:145-161. DOI: 10.1016/j.jclepro.2012.07.001

[13] Guidelines accompanying Commission Delegated Regulation (EU) No 244/2012 of 16 January 2012 supplementing Directive 2010/31/EU of the European Parliament and of the Council on the energy performance of buildings by establishing a comparative methodology framework for calculating cost-optimal levels of minimum energy performance requirements for buildings and building elements (2012/C 115/01), Official Journal of the European Union, C 115/1

[14] The Building Performance Institute Europe (BPIE). Implementing the cost-optimal methodology in EU countries – lessons learned from three case studies [Internet]. 2013. Available from: https://www.bpie.eu/publication/implementing-the-cost-optimal-methodology-in-eu-countries/ [Accessed: March 12, 2019]

[15] EN 15459:2010. Economic evaluation procedure for energy systems in buildings

[16] Statistical Office of the Republic of Slovenia (SURS) [Internet]. Available from: http://www.stat.si/StatWeb/pregled-podrocja?idp=30&headerbar=4 [Accessed: July 4, 2020]

[17] Humbert S, et al. IMPACT 2002+: User Guide [Internet]. Quantis, Lausanne, Switzerland. 2012. Available from: http://www.quantis-intl.com/pdf/IMPACT2002_UserGuide_for_vQ2.21.pdf [Accessed: July 10, 2020]

[18] Humbert S, et al. IMPACT 2002+ vQ2.22: Characterization factors [Internet]. 2014. Available from: http://www.quantis-intl.com/impact2002.php [Accessed: April 10, 2020]

[19] Goedkoop M et al. ReCiPe 2008. A Life Cycle Impact Assessment Method Which Comprises Harmonised Category Indicators at the Midpoint and the Endpoint Level. Netherlands: Ministrie van Volkshuisvesting, Ruimtelijke Ordening en Milieubeheeer; 2009

[20] PURES. Pravilnik o učinkoviti rabi energije v stavbah (Rules on efficient use of energy in buildings with a technical guideline). Ljubljana: Uradni list RS, št. 52/10 in 61/17 – GZ); 2010

Section 3

New Approaches

Chapter 8

Introduction of ZEB Technology in Japan

Jihui Yuan

Abstract

It is necessary to reduce energy consumption in order to combat global warming and stabilize energy supply and demand. In particular, final energy consumption in the business sector (buildings such as office buildings and commercial facilities) accounted for about 16.1% of Japan's total in FY2018 database, an increase from about 12.6% in FY1990 database. Therefore, there is a need for the spread of zero-energy building (ZEB), which can significantly reduce the energy consumption in buildings. Since people are active in the building, energy consumption cannot be completely reduced to zero; however, it can be closer to ZEB by reducing the energy used in the building and creating energy in the building as much as possible. This chapter introduces some technologies of energy saving and energy creation to realize ZEB in general buildings in Japan.

Keywords: global warming, zero-energy building, energy saving, passive technology, active technology, creative technology

1. Introduction

A zero-energy building (ZEB), which is also known as a net ZEB, is a building with net zero-energy consumption, meaning the total amount of energy used by the building on an annual basis is equal to the amount of renewable energy created on the site or in other definitions by renewable energy sources offsite, using technologies such as heat pumps, high efficiency windows and insulation, and solar panels [1, 2]. In general, ZEB is a building that aims to reduce the annual balance of primary energy consumed by the building while realizing a comfortable indoor environment. Since people are active in the building, energy consumption cannot be completely reduced to zero; however, it can be set to zero by the means of energy saving and energy creation (as shown in **Figure 1**). Achieving zero energy is an ambitious yet increasingly achievable goal that is gaining momentum across geographic regions and markets. Private commercial property owners have a growing interest in developing ZEBs to meet their corporate goals, and in response to regulatory mandates, federal government agencies and many state and local governments are beginning to move toward ZEB targets [3].

The introduction of ZEBs makes buildings more energy efficient and reduces the rate of carbon emissions once the building is in operation; however, there is still a lot of pollution associated with embodied carbon of buildings [4]. The importance

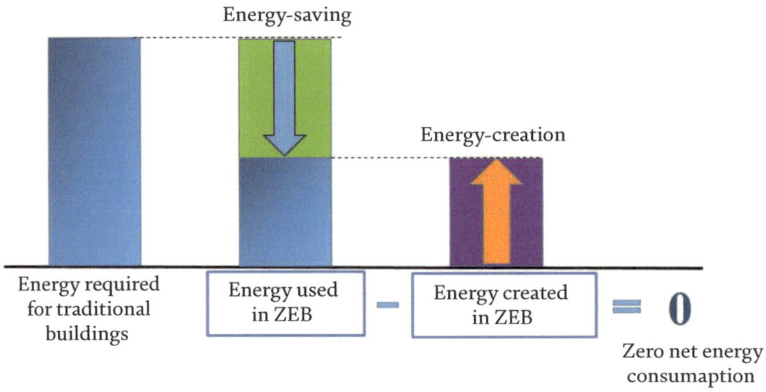

Figure 1.
Conceptual diagram for realizing zero-energy building.

of embodied carbon will grow as it will begin to account for the greater portion of a building's carbon emissions. In some newer and energy efficient buildings, embodied carbon has risen to approximately 47% of the building's lifetime emissions. Focusing on the embodied carbon is part of optimizing construction for climate impact and zero carbon emissions require slightly different considerations from optimizing only for energy efficiency [5]. One way to reduce the embodied carbon is by using low-carbon materials for construction such as straw, wood, linoleum, or cedar. A study reported that for materials such as concrete and steel, options to reduce embodied emissions do exist; however, these are unlikely to be available at large scale in the short term [6].

The ZEBs harvest available energy to meet their electricity and heating or cooling needs. The most common way to harvest energy is to use roof-mounted solar photovoltaic (PV) panels that can turn the solar radiation into electricity. Other common way such as using heat pumps, which can harvest heat and cool from the air (air-sourced) or ground near the building (ground-sourced otherwise known as geothermal), is also adopted to create energy. In the case of individual houses, various microgeneration technologies may be used to provide heat and electricity to the building, by using solar cells or wind turbines for electricity, and biofuels or solar thermal collectors linked to a seasonal thermal energy storage (STES) for space heating. The STES can also be used for summer cooling by storing the cold of winter underground. To cope with fluctuations in demand, ZEBs are frequently connected to the electricity grid, and they export electricity to the grid when there is a surplus and draw electricity from the grid when not enough electricity is being produced.

The most cost-effective steps toward a reduction in a building's energy consumption usually occur during the design process [7]. To achieve efficient energy use, zero-energy design departs significantly from conventional construction practice. The successful ZEB designers typically combine time-tested passive solar, or artificial/fake conditioning, principles that work with the on-site assets. Sunlight and solar heat, prevailing breezes, and the cool of the earth below a building can provide daylighting and stable indoor temperatures with minimum mechanical means. ZEBs are normally optimized to use passive solar heat gain and shading, combined with thermal mass to stabilize diurnal temperature variations throughout the day, and in most climates are super-insulated [8].

Countries around the world have been gradually implementing different policies to tackle ZEB, as a response to global warming and increasing greenhouse gas

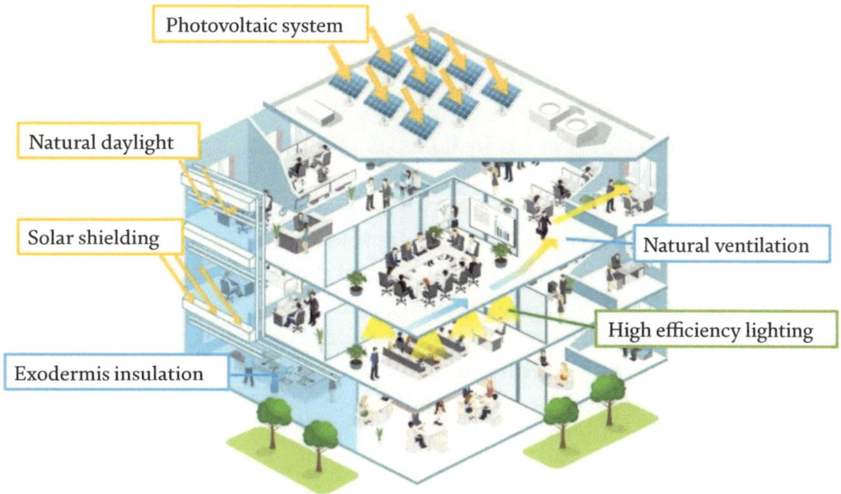

Figure 2.
Conceptual diagram of a general building to which energy-saving technology is applied.

emissions. In 2015, the Paris Agreement was created under the United Nations Framework Convention on Climate Change (UNFCC) with the intent of keeping the global temperature rise of the twenty-first century below 2°C and limiting temperature increase to 1.5°C by limiting greenhouse gas emissions [9]. While there was no enforced compliance, 197 countries signed the international treaty which bound developed countries legally through a mutual cooperation where each party would update its Intended Nationally Determined Contributions (INDC) every 5 years and report annually to the conference of the parties (COP) [10]. Due to the advantages of energy efficiency and carbon emission reduction, ZEBs are widely being implemented in many different countries as a solution to energy and environmental problems within the infrastructure sector [11].

After the March 2011, Fukushima earthquake followed by the up with Fukushima Daiichi nuclear disaster, and Japan experienced severe power crisis that led to the awareness of the importance of energy conservation. In 2012, Ministry of Economy, Trade and Industry (METI), Ministry of Land, Infrastructure, Transport and Tourism (MLIT) and Ministry of the Environment of Japan summarized the road map for low-carbon society, which contains the goal of ZEB to be standard of new construction in 2020 [12]. This chapter introduces the technologies to realize ZEB in general buildings (as shown in **Figure 2**) in Japan, in terms of i) passive technologies (solar shielding, exterior skin insulation, natural daylight, and natural ventilation), ii) active technologies (high-efficiency lighting, high-efficiency air conditioning, etc.), and iii) creation technologies (photovoltaic power generation, biomass power generation, etc.).

2. Passive technologies

Passive technology is a technology for reducing the amount of required energy or energy demand to properly maintain the environment inside a building. It includes solar shielding, exterior skin insulation (walls and windows), natural daylight utilization, and natural ventilation.

2.1 Solar shielding

Solar shielding is a technology that shields the sunlight that enters through the roof, exterior walls, and windows, and suppresses the cooling load. Particularly in the summer, once solar radiation (or heat) enters the room, a large amount of cooling energy is consumed to cool the heat; thus, solar shielding is considered as an important technology to realize a comfortable indoor environment. On the other hand, in winter, it is better to take in a considerable amount of solar heat to reduce the heating load. There is also a need to take in natural daylight well even in the summer from the viewpoint of reducing lighting energy. In this way, it is necessary to consider measures to meet the conflicting performance requirements, such as suppressing the intrusion of solar heat during cooling in the summer and taking in the sunlight during heating in the winter. As shown in **Figure 3**, specific measures to block sunlight at openings include blinds, louvers, eaves, and high-performance glass. By effectively combining these measures, it is possible to successfully cope with multiple conflicting performance requirements as described above.

The blinds are intended to prevent direct sunlight into the room, but they are also expected to work as a daylighting system that takes in natural light as indoor lighting. Recently, a "gradation blind" that efficiently takes in natural light into the room, while preventing the invasion of solar heat by optimally controlling the angle of the blind slats one by one and reflecting the light, has been developed.

The eaves are basically immovable, but by properly designing their length based on the solar altitude and the height of the windows, the sunlight with high solar altitude in summer can be blocked and the sunlight with low solar altitude in winter can be taken in.

For walls and other areas other than openings, the use of plants and materials with high solar reflectance can improve thermal insulation and solar radiation reflectance, thereby improving solar shielding performance.

2.2 Exterior skin insulation

The exterior skin insulation technology includes the use of high-performance exterior wall insulation and high-performance insulating and thermal barrier windows. High-performance exterior wall insulation can reduce the amount of energy required to maintain a comfortable indoor temperature compared to a non-insulated building by controlling the flow of heat in and out of the building by constructing

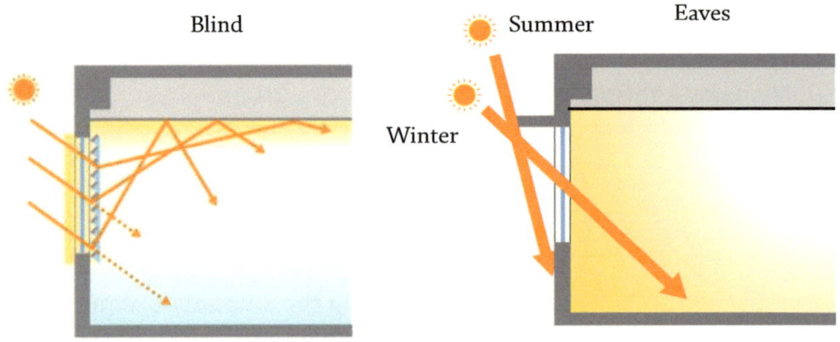

Figure 3.
Specific measures for solar shielding at openings.

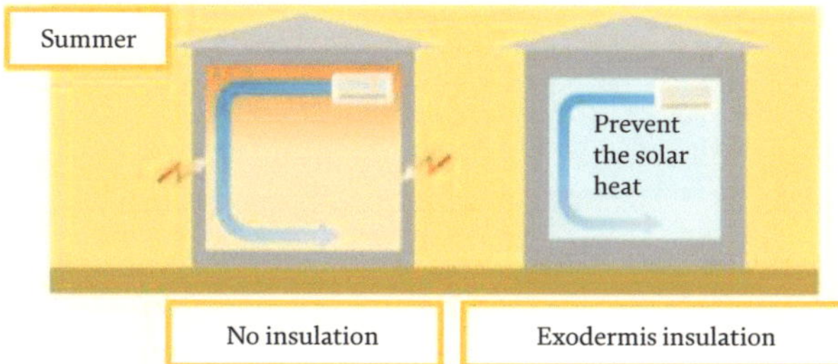

Figure 4.
Image of the effect of insulation in summer.

the exterior skin (roof, walls, floor, etc.). By preventing the penetration of solar heat in summer (**Figure 4**) and the escape of indoor heat in winter (**Figure 5**), energy consumption efficiency for heating and cooling can be improved, and the difference between the surface temperature of the building frame and the room temperature can be reduced, thereby minimizing temperature differences and unevenness in the room.

There are two main types of insulation materials: fiber-based and foam-based, both of which make use of the properties of gases (such as air) that make it difficult to transfer heat. Fiber-based insulation secures heat insulation by retaining air in the gaps between the fine fibers. Foam-based insulation also achieve high-heat insulation properties by trapping air or gases with higher-heat insulation properties inside the bubbles.

Nevertheless, not all regions need to use the same insulation design. A research proposed to optimize the combination of surface reflectivity and the insulation thickness of exterior walls for energy savings in regions of Japan [13]. Calculations of building thermal loads and economic analysis of the total cost for six cities from high-latitude to low-latitude regions of Japan were carried out for a range of surface reflectivity and insulation thickness of exterior walls, and the optimum surface solar reflectivity and insulation thickness for each region were proposed.

Since the openings of a building have the highest heat input and output of the exterior skin, it is important to control the heat input and output by adopting windows with glass that has high thermal insulation performance.

One of the most common types of window glass with high thermal insulation performance is double glazing (as shown in **Figure 6**). In general, double glazing glass has a hollow layer between two panes of glass, which is filled with dry air with

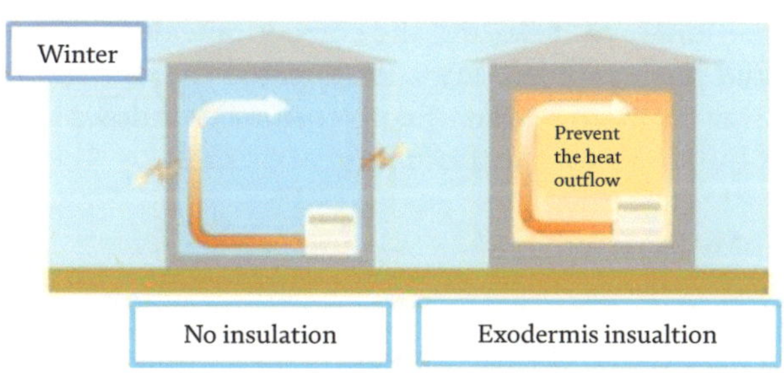

Figure 5.
Image of the effect of insulation in winter.

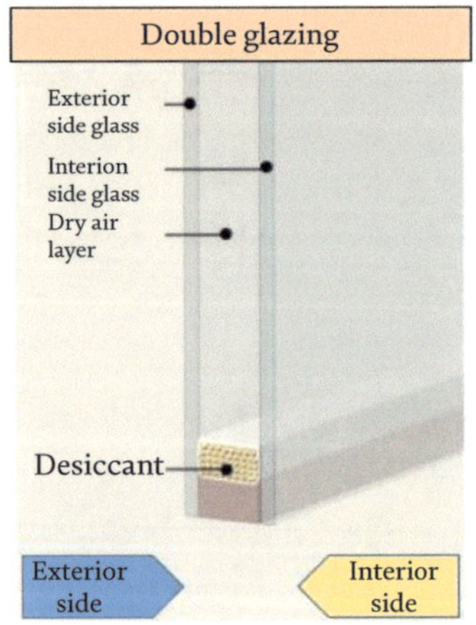

Figure 6.
Common type of double glazing window glass with high thermal insulation performance.

low thermal conductivity or argon or krypton gas with lower thermal conductivity, to improve thermal insulation performance. The thicker the hollow layer, the better the thermal insulation performance, but if the layer exceeds 16 mm, the thermal resistance does not increase due to air convection.

As shown in **Figure** 7, low-E double glazing glass is double glazing glass coated with a special metal film (low-E film) such as tin oxide or silver. This special metal film makes it more difficult for thermal radiation in the hollow layer between the glasses to be transmitted. In summer, solar radiation energy incident on plate glass is reflected outside the room. In winter, it reflects heating heat indoors. As a result, thermal insulation and heat shielding performance can be further enhanced. In summary, by coating the inside of the glass on the indoor side, it is possible to improve the thermal insulation performance to prevent heat from flowing out of the room, and by coating the inside of the glass on the outdoor side, it is possible to improve the thermal barrier performance to prevent solar heat from flowing in.

The thermal transmittance of a single pane of glass is about 5.0 to 6.0 W/m^2K, whereas it is about 1.8 to 3.3 W/m^2K for double glazing and about 0.76 to 2.6 W/m^2K for low-E double glazing.

For this technology, a Japanese study proposed the thermal performance values for conventional windows and low-E windows that include air-flow windows and push-pull windows [14]. The solar heat gain coefficients together with the transmittances, the overall coefficients of heat transfer, and the long-wave radiation factors were presented for conventional windows and the correction values were presented for air-flow windows and push-pull windows. It indicated that the thermal performance of low-E windows was better than that of conventional windows and leads to reduce the thermal load of buildings for the ultimate goal of ZEB.

2.3 Natural daylight utilization

The natural daylight utilization is a technology to reduce energy consumption by bringing in natural daytime light (daylight) through building openings to brighten the room and reduce the use of artificial lighting (or indoor lighting). The energy consumption used for artificial lighting is second only to that used for air-conditioning

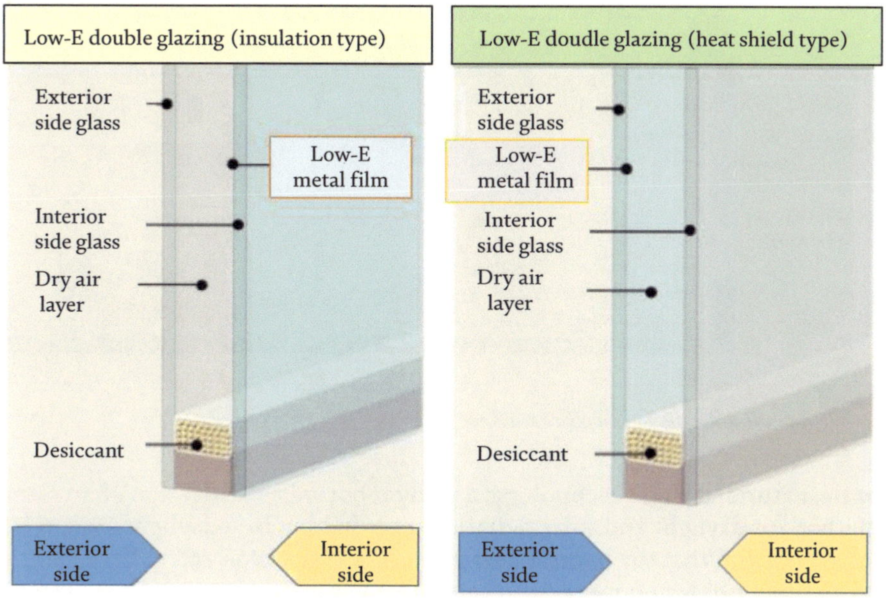

Figure 7.
Low-E double glazing glass coated with a special metal film, the left insulation type is often used in winter and the right heat shield type is often used in summer.

in a typical office building, making it a major source of energy for the building. If daylighting can provide the necessary brightness in a room, energy consumption can be reduced by turning off or dimming the lights to reduce the amount of light, which can also lead to greater energy independence.

As shown in **Figures 8** and **9**, there are two methods of natural lighting: One is to let daylight in directly through the openings of the building to secure the brightness of the room, and the other is to set up a stairwell, a parapet, or reflective eaves to guide the light deeper into the room.

The former (**Figure 8**) is the technique of installing an opening (top light) at the top of a space, etc. For the latter (**Figure 9**), there is a method of installing a "light shelf" in the middle of the window surface of the building to reflect sunlight on the upper surface and bring more light into the interior ceiling to brighten the room. There is also a method called a "light duct system."

Figure 8.
Example of top light utilization in a building.

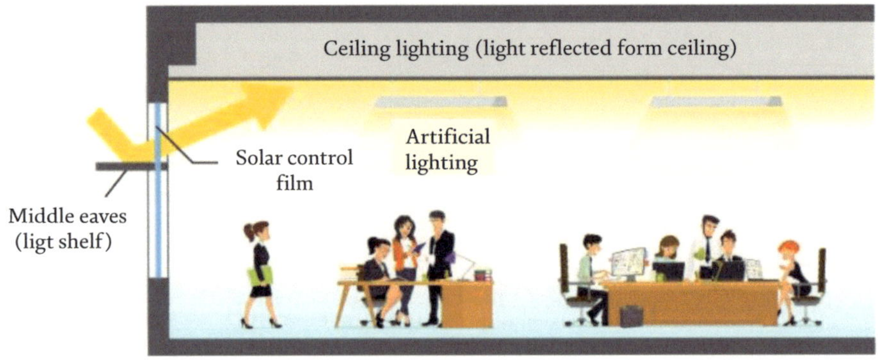

Figure 9.
Example of light shelf utilization in an office building.

For the natural daylight technology, a study reported a verification of an estimation method for daylight and solar radiation introduction by a daylight system [15]. The results showed that the natural daylight utilization is required to have two contradictory functions for energy conservation: reducing lighting energy by introducing daylight and reducing air-conditioning load by shielding from sunlight. In particular, daylight-using facades need to be planned with a balance between the two functions of daylight introduction and solar radiation shielding.

Daylight has the following characteristics: It changes with time, it may bring more brightness than necessary for the indoor visual environment, and it is accompanied by heat. Therefore, when using daylighting, it is important to consider these characteristics and adopt a lighting method that is appropriate for the space characteristics and usage. Without natural lighting suitable for the space characteristics and usage, energy conservation may not be achieved because building users may block daylight, or the energy consumption for cooling may increase more than the reduction in lighting power.

2.4 Natural ventilation

As shown in **Figure 10**, it creates a wind path through the building to enable natural ventilation and natural airflow. This technology is mostly applied to mid-rise offices; however, we can also propose an optimal natural ventilation system for high-rise offices by predicting the wind flow and wind pressure on the exterior walls such as double-skin facade.

Double skin is a system with two skins (glass surfaces) that ventilate the inside of the two skins with outside air. In the summer, the blinds inside the double skin are lowered to shield the building from the sun's rays, and since the sun's rays become heat, the heat is removed by ventilating the building with outside air, thereby reducing the cooling load. In winter, the double-skin ventilation is stopped and the double or triple glazing improves the thermal insulation performance. In the middle of the year, stable natural ventilation is possible through the double skin.

In order to understand the performance of a natural ventilation system that combines solar chimneys and underground pits installed in a university building in Kitakyushu City, Japan, a study conducted a measurement survey over a period of 4 years after the opening of the school [16]. The results indicated that the designed natural ventilation system can greatly reduce the energy consumption of air

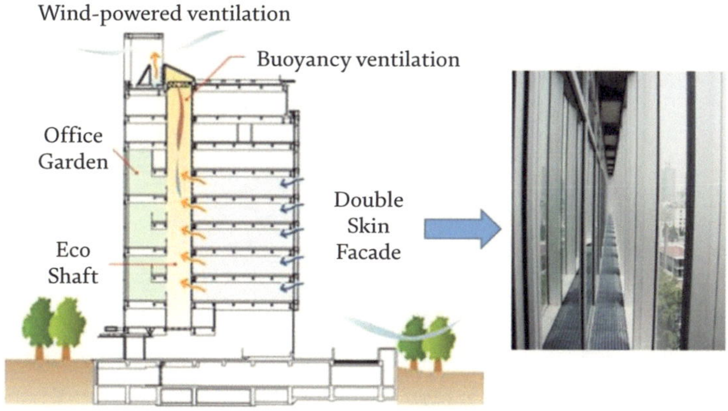

Figure 10.
Example of natural ventilation design in an office building.

conditioning in cooling period, and it can also provide a more comfortable indoor thermal environment.

3. Active technologies

Active technology is a technology for efficient use of energy. It includes high-efficiency lighting, high-efficiency air conditioning, etc. In this section, we briefly introduce these two methods of efficient energy use.

3.1 High-efficiency lighting

In order to reduce lighting energy consumption, it is important to actively use daylighting, for example, by adopting natural lighting techniques. At the same time, it is possible to reduce energy consumption while providing an appropriate lighting environment (illuminance, etc.) by using more efficient lighting equipment such as LED lighting to compensate for the lack of brightness from the use of daylight alone. In addition, by properly controlling such lighting equipment, even higher energy-saving effects can be expected.

Figure 11 shows the examples of lighting control by human sensors and wireless remote thermosat, task-ambient lighting control, and a combination of these controls.

As shown in **Figure 11** on the left, this control uses human sensors to detect the presence or absence of people and turns on or off the air conditioning and lighting. In addition, a remote thermostat that accurately measures the temperature of the area where the person is and efficiently controls the air conditioning.

As shown in **Figure 11** on the right, the ceiling lighting should function as ambient lighting for room ambiance, while desk brightness is adequately provided by the task lighting for work at hand. This is the most effective method for reducing lighting power.

For the high-efficiency lighting utilization in Japan, a study showed that when the high-efficiency lighting fixtures and lighting control systems have been introduced in an office buildings of Japan, the electricity consumption savings of 30 to 50% per year can be expected, and a significant energy savings can be achieved without degrading the quality of lighting [17].

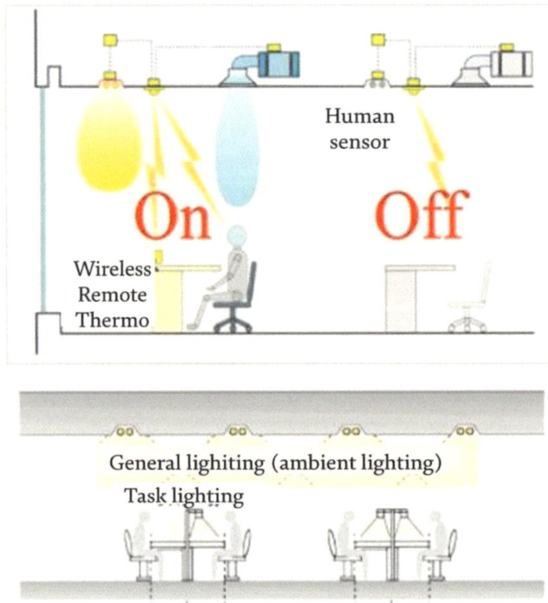

Figure 11.
Example of optimal lighting system control in an office building.

3.2 High-efficiency air conditioning

In order to reduce the energy consumption of air conditioning, it is important to control the load of heating and cooling by adopting passive technologies such as external skin insulation and solar radiation shielding that are elaborated in Section 2. However, since it is often difficult to maintain a comfortable indoor environment with these measures alone, it is important to reduce energy consumption while maintaining a comfortable thermal environment by using an air-conditioning system with higher efficiency and appropriate control to compensate for this. In a typical office building, energy consumption by the air-conditioning system accounts for the largest percentage of the total energy consumption, and the importance of reducing it is very high.

Air-conditioning systems can be broadly divided into central heat source systems and individual distributed heat source systems. In the central heat source system (as shown in **Figure 12**), heat sources are concentrated in machine rooms, etc., and cold and hot water is pumped to the air conditioner for air conditioning. In the individual distributed heat source system (as shown in **Figure 13**), heat sources are distributed and transported using refrigerant piping to air condition for each floor or zone. Both the central heat source system and the individual distributed heat source system consist of "heat source equipment," "heat transfer equipment," and "air conditioner equipment". In the case of a distributed heat source system, the heat source and air-conditioning equipment are integrated into a single unit. Therefore, energy consumption can be reduced by adopting more efficient equipment and implementing appropriate controls for each facility.

In general, the central heat source system is used in large buildings, while the individual distributed heat source system is used in many small buildings.

Measures to reduce the energy consumption of air-conditioning equipment include air-conditioning systems that separate latent heat from sensible heat to adjust

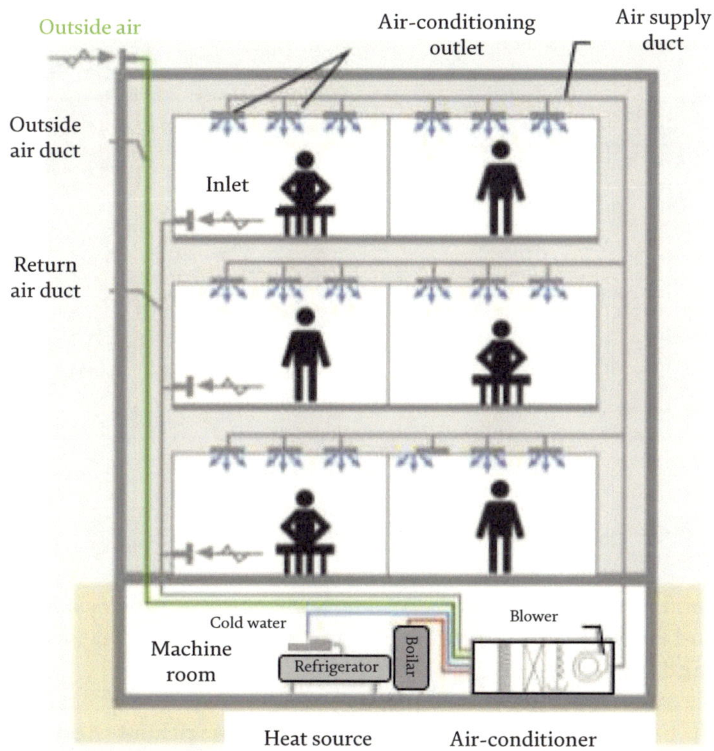

Figure 12.
Example of the central heat source system.

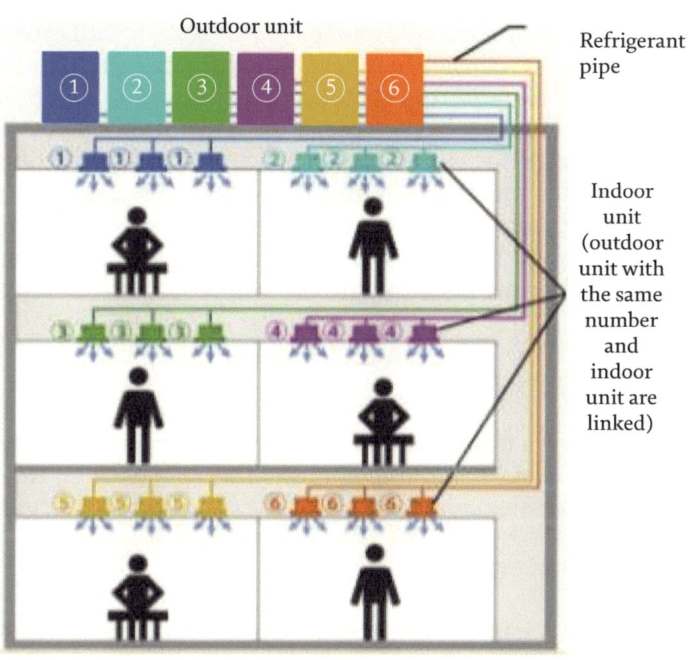

Figure 13.
Example of the individual distributed heat source system.

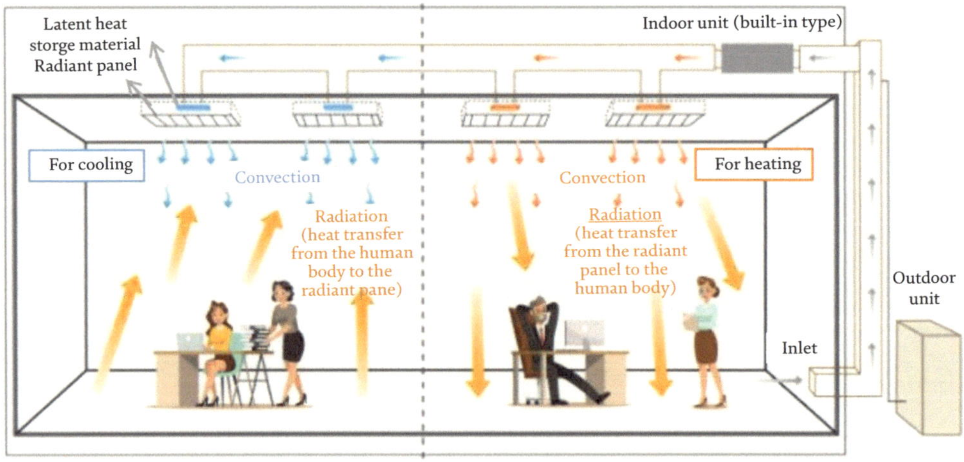

Figure 14.
Conceptual diagram of radiant heating and cooling air-conditioning systems.

temperature and humidity separately, and new air-conditioning systems such as radiant heating and cooling air-conditioning systems (**Figure 14**) that focus not only on temperature and humidity but also on the comfort felt by people.

A radiant heating and cooling system is a system that uses the effect of "radiation (the transfer of heat from a higher to a lower temperature without the use of materials)" to adjust the experience of building users, thereby easing the indoor set temperature and saving energy. Compared to conventional air conditioner systems, this system is more comfortable and less uncomfortable due to airflow drafts and uneven temperatures.

For the high-efficiency air-conditioning system, a study showed that a kind of developed high-efficiency air-conditioning control system with promotion of energy-saving behaviors were installed in may stores of Japan [18]. The "promotion of energy-saving behavior" supports the voluntary establishment of employee behavior by proposing optimal energy-saving behavior for each store based on AI power prediction and displaying screens using nudge theory *via* tablets distributed to stores. The high-efficiency air-conditioning control system suppresses demand based on AI power prediction and operates the air-conditioning compressor at a high COP load rang, and achieves energy savings of more than 3% through behavioral promotion and more than 4% through air-conditioning control.

4. Creation technologies

Creation technology is a technology to use renewable energy to create energy. It includes photovoltaic power generation, biomass power generation, etc. In this section, we briefly introduce these two methods of the creation technologies.

4.1 Photovoltaic power generation

A photovoltaic power generation system generally refers to a power generation system that uses semiconductors to convert light energy from the sun into electrical energy and consists of solar cell modules and arrays, junction boxes and collectors, and power conditioners, as shown in the **Figure 15**.

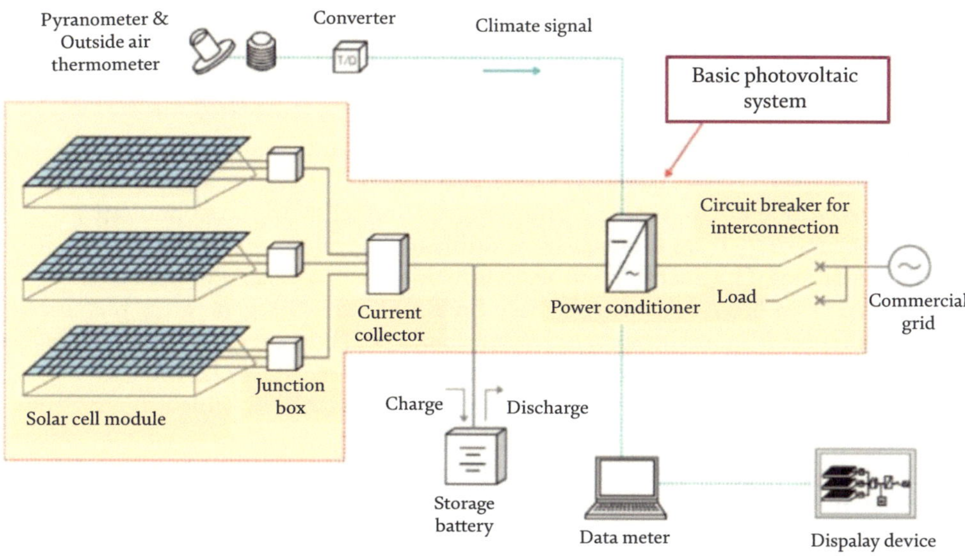

Figure 15.
Conceptual diagram of a photovoltaic power generation system.

The benefits of light energy from the sun are available everywhere, although they vary from region to region, making it the most versatile of all energy creation technologies. When installed on the rooftops of office buildings and commercial facilities, the electricity generated by the photovoltaic power system can be used to meet part of the electricity demand, since the working hours and business hours coincide with the power generation hours. In addition, the system can be promoted as a part of environmental conservation activities, which is a social responsibility of a company, and is beneficial in raising the environmental awareness of employees and securing power in times of disaster.

Particularly in the case of low-rise buildings, the rooftop area is large in relation to the size of the building, and a reasonable amount of power can be expected to be generated in relation to the power demand. On the other hand, in the case of high-rise buildings, since the rooftop area is small compared to the size of the building, the amount of electricity generated by the photovoltaic power system will be small in relation to the electricity demand. Recently, however, there has been progress in the development of "building-integrated photovoltaic systems" that can be installed not only on the rooftops of buildings, but also on walls and windows.

For the situation of photovoltaic power generation in Japan, its research and development and widespread use in Japan began with the oil crisis of the 1970s [19]. At the time, research and development were focused on the use of solar power as an alternative energy source that did not consume petroleum fuel. Later, the movement was further accelerated by global environmental issues and global warming prevention in the 1990s. In the 1990s, the first residential photovoltaic power generation systems were commercialized, and photovoltaic power generation systems evolved from being mainly used for research and development and special purposes to supplying electricity to the general public. In Japan, the feed-in tariff system for renewable energy started in July 2012. In July 2015, the government's Committee on Energy Supply and Demand (CESD) issued a report on the electricity market and presented a supply and demand forecast for various power sources for 2030. In this energy mix, photovoltaic power generation will be responsible for supplying 7% of the electricity demand in 2030.

4.2 Biomass power generation

As shown in **Figure 16**, biomass power generation refers to the technology of generating electricity from biomass (renewable biological resources) such as wood and plant residues. The energy obtained from biomass is also called biomass energy.

When biomass is burned, as with fossil fuels, CO_2 is always generated, but since plants absorb the CO_2 and grow to reproduce biomass, the total amount of CO_2 in the atmosphere does not increase (carbon neutral as shown in **Figure 17**).

By combining this system with photovoltaic power generation, which changes the amount of electricity generated depending on the weather and time of day, it is expected that renewable energy will be supplied in accordance with the demand for electricity.

For the situation of biomass power generation in Japan [20], the Biomass Power Producers Association (BPPA) was established in late 2016, with the aim of addressing the concerns of power producers and promoting the healthy development of biomass power producers. Biomass power generation capacity in Japan reached approximately 3.0 GW by the end of FY2016, approximately 4.0GW by the end of FY2017, and since then, it application has been steadily growing in the years.

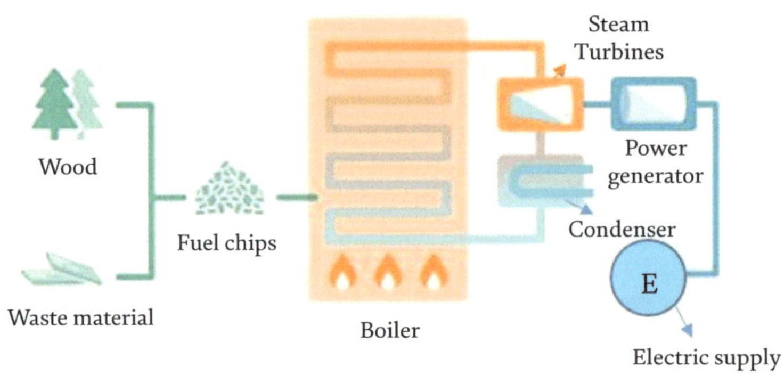

Figure 16.
Conceptual diagram of how biomass power generation works.

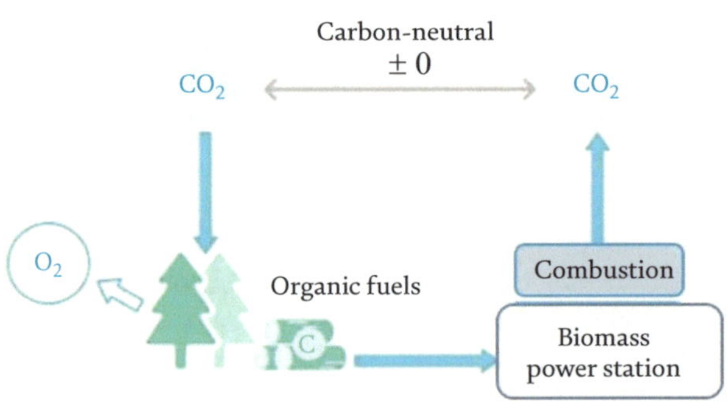

Figure 17.
Carbon neutral of biomass power generation system.

5. Summary and conclusions

This chapter provides an overview of passive, active, and creative energy technologies that are particularly important for realizing ZEB in buildings commonly used in Japan.

The main knowledge obtained in this chapter is summarized as follows:

- Solar shielding technologies that include blinds and eaves installed in the openings can shield the sunlight to enter through windows, and suppress the cooling load in the summer. However, in winter, it is better to take in a considerable amount of solar heat to reduce the heating load by appropriate opening design.

- Exterior skin insulation technologies that include high-performance exterior wall insulation and low-E windows can reduce the amount of energy required to maintain a comfortable indoor temperature compared to a non-insulated building exterior walls and windows by controlling the flow of heat in and out of the building by constructing the exterior skin.

- Natural daylight utilization is required to have two contradictory functions for energy conservation: reducing lighting energy by introducing daylight and reducing air-conditioning load by shielding from sunlight, through proper design of daylight utilization openings.

- A proper natural ventilation system designed in buildings can greatly reduce the energy consumption of air conditioning in the summer cooling period, and it can also provide a more comfortable indoor thermal environment.

- High-efficiency lighting utilization can save the electricity consumption without degrading the quality of lighting.

- High-efficiency air conditioning can reduce energy consumption while maintaining a comfortable thermal environment.

- As a renewable green energy source, the photovoltaic power generation and biomass power generation technologies have been widely applied in Japan and have been steadily growing in the years.

To actually realize ZEB, it is important to consider the following steps: (i) reduce energy demand through passive technology, (ii) use energy without waste through active technology for the demand that is absolutely necessary, and (iii) provide energy through energy creation technology.

Moreover, in the operational phase of a building, it is also important to have energy management technology to determine where energy waste is occurring and how to efficiently operate the facilities. This energy management technology will help reduce energy consumption on an ongoing basis.

Conflict of interest

The authors declare no conflict of interest.

Author details

Jihui Yuan
Department of Architecture and Civil Engineering, Graduate School of Engineering, Toyohashi University of Technology, Toyohashi, Aichi, Japan

*Address all correspondence to: yuan@ace.tut.ac.jp

References

[1] Torcellini P, Pless S, Deru M, Crawley D. Zero energy buildings: A critical look at the definition. In: Conference Paper NREL/CP-550-39833. California: National Renewable Energy Laboratory; August 14–18, 2006. pp. 1-12

[2] US Department of Energy. A Common Definition for Zero Energy Buildings. Energy Efficiency & Renewable Energy, DOE/EE-1247, September 2015. Available from: https://www.energy.gov/sites/prod/files/2015/09/f26/A%20Common%20Definition%20for%20Zero%20Energy%20Buildings.pdf [Accessed: November 23, 2021]

[3] Energy Government. Zero Energy Buildings. Available from: https://www.energy.gov/eere/buildings/zero-energy-buildings [Accessed: October 3, 2021]

[4] Alter L. Architecture 2030 Goes After Embodied Carbon and This Is A Very Big Deal. TreeHugger. Retrieved 2019-12-08. Available from: https://www.treehugger.com/architecture-goes-after-embodied-carbon-and-very-big-deal-4852211 [Accessed: October 3, 2021]

[5] Alter L. The New Carbon Architecture, Or Why We Should Be "Building Out of Sky" (Book Review). TreeHugger. Retrieved 2019-12-08. Available from: https://www.treehugger.com/new-carbon-architecture-or-why-we-should-be-building-out-sky-4853771 [Accessed: October 3, 2021]

[6] Ebrahimi G. Embodied Carbon and Deep Retrofits. Canada: Pembina Institute; 2020. Available from: http://www.jstor.org/stable/resrep25464

[7] Vieira R. The Energy Policy Pyramid-A Hierarchal Tool For Decision Makers. Fifteenth Symposium on Improving Building Systems in Hot and Humid Climates, July 24-26, 2006, Orlando, FL. Available from: http://www.fsec.ucf.edu/en/publications/html/FSEC-PF-401-06/ [Accessed: October 3, 2021]

[8] Frej A, editor. Green Office Buildings: A Practical Guide to Development. Washington, DC: Urban Land Institute; 2005. pp. 138-142. ISBN 978-0-87420-937-2

[9] The Paris Agreement. United Nations Framework Convention on Climate Change. Retrieved 2020-11-05

[10] Paris Agreement | Summary & Facts. Encyclopedia Britannica. Retrieved 2020-11-05

[11] Liu Z, Zhou Q, Tian Z, He B, Jin G. A comprehensive analysis on definitions, development, and policies of nearly zero energy buildings in China. Renewable and Sustainable Energy Reviews. 2019;**114**:109314. DOI: 10.1016/j.rser.2019.109314

[12] Reform Lab. What is Zero-Energy Housing (ZEH). Available from: https://standard-project.net/smarthouse/zeh.html [Accessed: October 3, 2021]

[13] Yuan J, Farnham C, Emura K, Alam MA. Proposal for optimum combination of reflectivity and insulation thickness of building external walls for annual thermal load in Japan. Building and Environment. 2016;**103**:228-237

[14] Kohri K, Ishino H. A study on performance values of windows for thermal load calculation. J. Environ. Eng., AIJ. 2006;**600**:39-44. (in Japanese)

[15] Ito S, Kawashima M, Suzuki M, Kakegawa S, Miura N, Nakamura T. Study on building system simulation for zero energy building (Part 1) verification of an estimation method for daylight and solar radiation introduction by a daylight system. Proceedings of the Japan Society of Air-Conditioning and Sanitary Engineering. 2014;**5**:205-208. (in Japanese)

[16] Shinada Y, Kimura K, Katsuragi H, Soog S. Field investigation on the performance of Natural ventilation system combined with solar chimney and underground pit for four years (Part 1) System operation and supply/exhaust air volumes of natural ventilation. The Society of Heating, Air-Conditioning Sanitary Engineers of Japan. 2009;**153**:45-56. (in Japanese)

[17] Sasai T. Lighting control using remote control switch. The Illuminating Society Journal. 1999;**83**(9):704-706. (in Japanese)

[18] Hayashi N. An energy management service that combines AI and nudge theory to promote energy-saving behavior and high-efficiency air conditioning control. Journal of the Japan Institute of Energy. 2021;**100**(3):309-313. (in Japanese)

[19] Kameda M. The Present State and the Perspective on the PV Power Generation. J. IEIE Jpn. 2016;**36**(6):363-366. (in Japanese)

[20] Kitamura S. Future of the Biomass Power Generation. Journal of the Japan Institute of Energy. 2018;**97**:100-104. (in Japanese)

Chapter 9

Adaptive Thermal Comfort of an Office for Energy Consumption-Famagusta Case

Halil Zafer Alibaba

Abstract

The aim of this study was to determine how much thermal comfort can be obtained through heat/energy transfers between the office/external air and the transparent/opaque surfaces of an office by combining different transparent and opaque wall surface ratios with different window opening percentages using dynamic thermal simulations. It found that the optimum window-to-wall ratio (WWR) for energy conservation is 40%, with a 20% window opening ratio. The 80% and 90% thermal comfort ranges of the adaptive thermal comfort methodology are found in May, October, September, and the yearly average, while June and August are only in the range of 80% acceptability. The office constantly loses heat through air flow with any glass size on its external facade and any window opening ratio. Moreover, all sizes of opaque and transparent internal surfaces transferred heat from outside by conduction, while the opaque wall similarly always transferred energy to heat up the office air internally and outside air externally through convection. The external glass also heats the office air by convection, except in the months of January, November, and December.

Keywords: passive heating and cooling, transparency ratio of the skin, adaptive thermal comfort, conduction, convection

1. Introduction

Natural ventilation has great potential for cooling buildings with a passive strategy because it improves user comfort and indoor air quality, while simultaneously reducing electricity usage demand [1–5]. Thermal comfort has a great impact on the well-being and performance of users, as well as the energy requirements of buildings [6]. Moreover, in developed countries, people spend 90% of their time indoors, which requires securing their well-being and a healthy environment [7].

Reducing active cooling requires passive solutions; therefore, utilizing glass on external walls has a great impact on thermal situations through its influence on energy demand for the cooling and heating of buildings [8].

An adaptive approach to thermal comfort relies on outdoor air temperature and the individuals' thermal environment [9]. Adaptive comfort limits are similar for hot-humid and hot-dry climates with a 0.6 coefficient value. Additionally, air movement is important for a hot-humid climate, while indoor relative humidity (RH) is important for a hot-dry climate [10]. In hot-humid climate environments, relative humidity is not significant when it is below 70%, but is really significant when higher than 70% with increasing air temperature [11].

Sustainable approaches for energy efficient buildings are very important to architects and engineers for the provision of comfortable and health-conscious buildings [12]. In addition to these, adaptive thermal comfort can be used to design energy efficient (low-energy) naturally ventilated buildings around the world [13].

The aim of this study was to determine the minimum, maximum, and yearly averages of naturally ventilated office performance, as well as the impact of the wall-to-window ratio on thermal comfort due to solar gain, heat loss, resultant temperature, relative humidity, and external air temperature. Moreover, the analysis focuses on the office's performance in relation to the winter and summer season (based on each month), heat transfer (conduction), and energy transfer (convection) for opaque and glass walls (internally and externally). In identifying the naturally ventilated office performance (minimum, maximum, and yearly) and the seasonal performance of the office (based on conduction-convection), this article hopes to fill in existing gaps in the literature for hot-humid climatic conditions.

2. Literature review

Heat transfer through the opaque walls of buildings is important for energy saving and thermal comfort issues. Cities in Turkey such as Ankara, Erzurum, Istanbul, and Izmir with different climates were analyzed using the TS825 standard (Turkish standard on thermal insulation requirements for buildings) [14]. It was found that 15-cm- and 25-cm-thick sandwich panels created a decrease of 65% and 80% heat loss and gain respectively during the worst winter and summer conditions. The optimal heat loss and gain ratios under different climatic conditions were determined using sandwich wall insulation. Heat transfer for different building orientations was found to be longer in the summer period due to solar radiation [15]. Radiative heat transfer was higher during daytime than the evening in summer, with no significant changes during the spring, autumn, and winter periods [16].

In the subtropical climate of China, out of eight free-running dormitories, 1829 returned questionnaires with subjective scales indicated that 15.9–28.2°C were acceptable temperatures, where 23.2°C was the preferred temperature and 22.1°C was the neutral temperature based on students' thermal perception and preferences. Moreover, the students' air movement sensation was 53% satisfied with the indoor air humidity when thermal sensation was neutral [17]. In Tuxtla Gutiérrez-México, 496 data points were collected from 27 educational buildings in the warm season. In air-conditioned mode, 48.1% of users felt comfortable, 44% felt cold, and 7.9% felt warmth. However, in naturally ventilated mode, 59.7% felt comfortable, 11% felt cold, and 29.3% felt warmth. Most of those who felt cold can have their thermal satisfaction improved by adapting rooms to slightly higher temperatures [18]. In the hot-humid part of China, naturally ventilated buildings have a thermal neutrality of 25.4°C, 23.5°C for 90% acceptability, and 27.4°C for the 80% acceptability range. In naturally ventilated buildings in China with a hot-humid climate, the Predicted Mean Vote (PMV) model can be used with a 0.822 expectancy factor [19].

In two Indian cities, Chennai with a humid climate and Hyderabad with a composite climate, the mean room temperature was 28.8°C for naturally ventilated mode but 26.2°C for air-conditioned mode with 45% and 48% relative humidity ratios, respectively [20].

Naturally ventilated buildings in hot-humid climate conditions should have a minimum of 0.65 m/s indoor air speed for thermal comfort [10]. In north-east Brazil, 90% acceptability of thermal comfort for a naturally ventilated building needs from 24 to 27°C room temperature with 0.4 m/s air movement, while from 27 to 29°C needs a minimum of 0.41–0.81 m/s air velocity, and a room temperature from 29 to 31°C needs a minimum of >0.81 m/s air velocity [21].

Tianjin is a city in China where 80% adaptive comfort acceptability ranges between room temperatures of 21°C and 27.3°C [22].

The national code for India has a narrow thermal comfort range that is between 21 and 26°C for all naturally ventilated building types and seasons. A questionnaire with 2610 responses found that the comfort ranges were 30.6°C with 0.30 clo dress and 0.62 m/s preferred air velocity for the summer season, and 25.2°C with 0.80 clo dress and 0.27 m/s preferred air velocity for the winter season [23].

A survey conducted in Spain found that 23.6°C is the observable average operative temperature in (free-running and air-conditioned) buildings. Moreover, it is very clear that an adaptive comfort model is suitable for hybrid buildings [24]. In India, the ideal comfort temperature was determined to be 27.3°C, while the actual preferred temperature is 24.5°C. Half of all the fans in offices start working after 31°C, with no fan needed up to 22.5°C [25].

3. Methodology

3.1 Adaptive thermal comfort method

Dynamic thermal simulations for this article were generated using the thermal analysis software EDSL Tas version 9.4.4 [26], which was simulated for Famagusta with hot and humid climatic conditions. The location of Famagusta can be seen in **Figure 1**. ASHRAE 55-2017 [28] standard for adaptive thermal comfort for 80% and 90% can

Figure 1.
Location of Famagusta on the map [27].

be observed in **Figure 2**. Heat transfer (conduction) and energy transfer (convection) opportunities were analyzed for the internal and external surfaces of both opaque and transparent surfaces in an office. Moreover, the office environment was that of a regular office (3 m by 5 m), including standard construction materials with one inlet and outlet function. The three-dimension (3D) and plan of the case study building can be seen in **Figure 3**. Furthermore, the weather file for Famagusta was also used for the simulations, as shown in **Figure 4**. Transparent surfaces on opaque walls and window opening percentages ranged from 10 to 100% each. A typical section of the case study building can be seen in **Figure 5** along with its yearly performances. In this chapter, all of the simulations used 0.5 ach of infiltration and 0 W/m^2 (lighting gain, occupancy/equipment gain) with 0.01 (CO_2)/hr/m^2 pollutant generations.

The opaque and transparent components of the case-study building are detailed in **Tables 1** and **2**. ASHRAE 55-2017 [28] was used to generate the acceptable thermal comfort conditions, shown in **Table 3** (80% and 90%), of a naturally ventilated office environment with minimum, maximum, and average yearly performances for solar gain (W), infiltration gain-heat lost (W), resultant temperature (°C), relative humidity (%), and external temperature (°C).

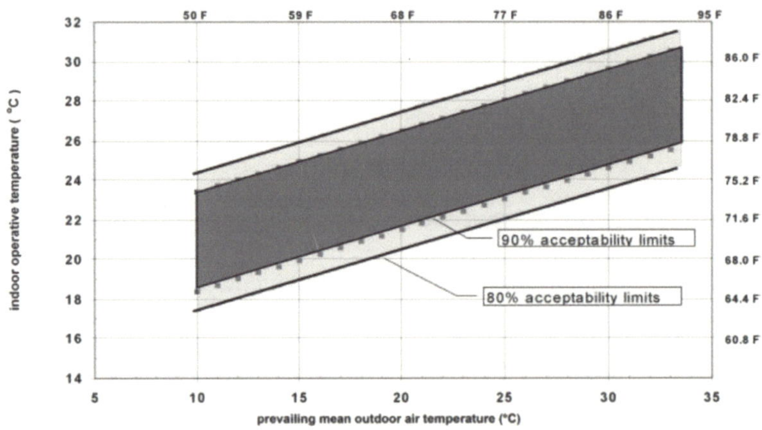

Figure 2.
ASHRAE 55-2017 standard on acceptable limits for the resultant temperature of a naturally ventilated building with met: 1.0–1.3 and 0.5–1.0 clo when the prevailing mean outdoor temperature is greater than 10°C and less than 33.5°C [28].

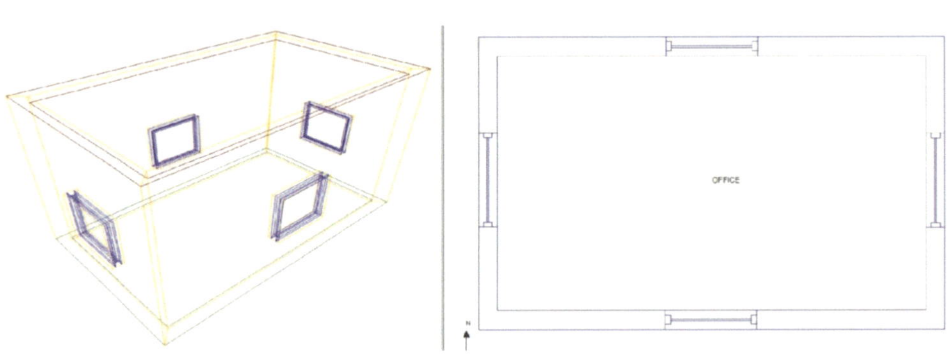

Figure 3.
The case study building.

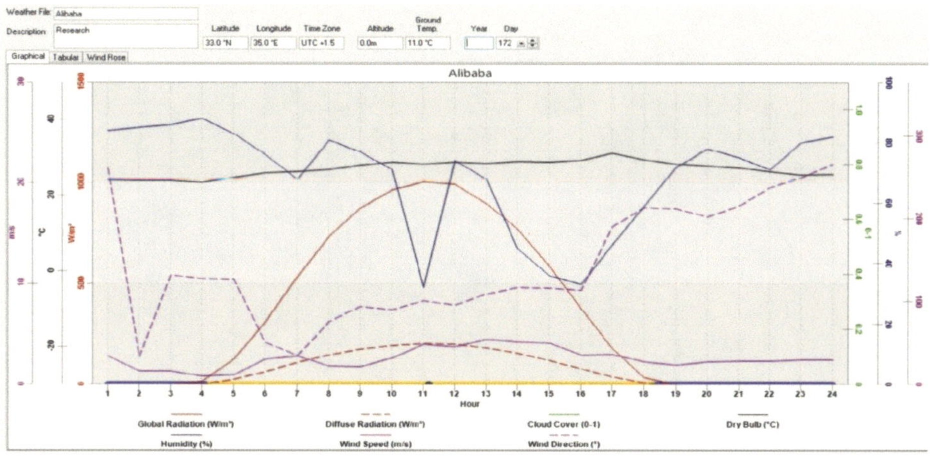

(a)

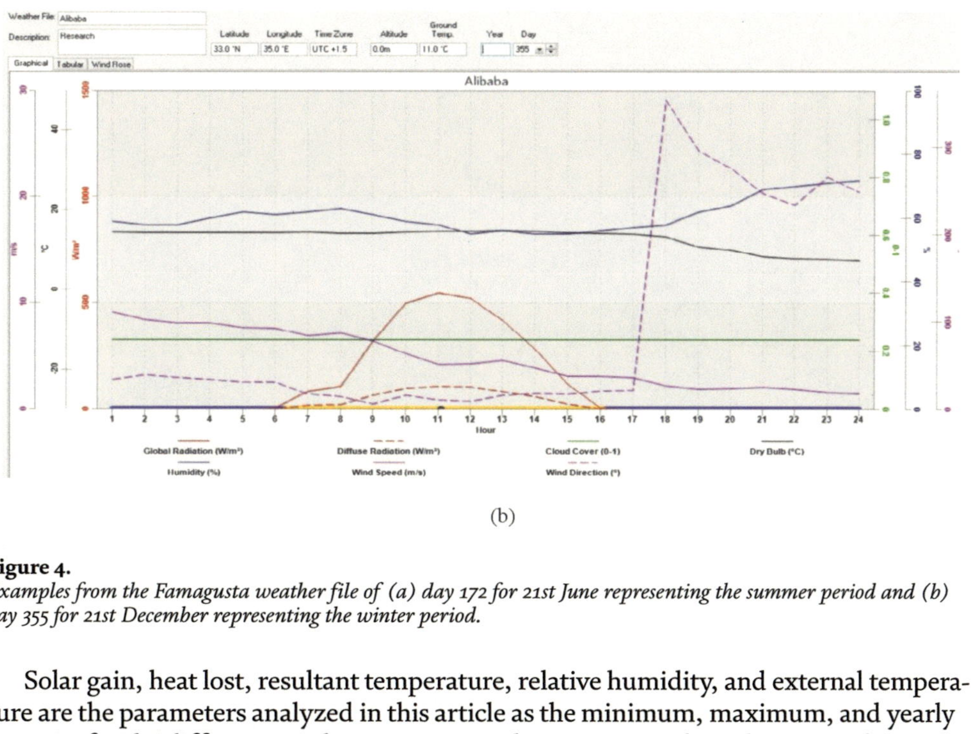

(b)

Figure 4.
Examples from the Famagusta weather file of (a) day 172 for 21st June representing the summer period and (b) day 355 for 21st December representing the winter period.

Solar gain, heat lost, resultant temperature, relative humidity, and external temperature are the parameters analyzed in this article as the minimum, maximum, and yearly averages for the different window openings and sizes. Seasonal conduction and convection performances of the studied office are based on monthly analysis, taken in conjunction with opaque/glass surface performances for internal/external surfaces. Global solar radiation (W/m^2), diffuse solar radiation (W/m^2), cloud cover (0–1), dry bulb temperature (°C), relative humidity (%), wind speed (m/s), and wind direction (°) are parameters used in the weather file of Famagusta for simulations as seen in **Figure 4**.

3.2 Inter-model validation of the article

An inter-model validation model for annual heat loss is used in this article because its numerical results are compared with previous results in the literature. Badeche and

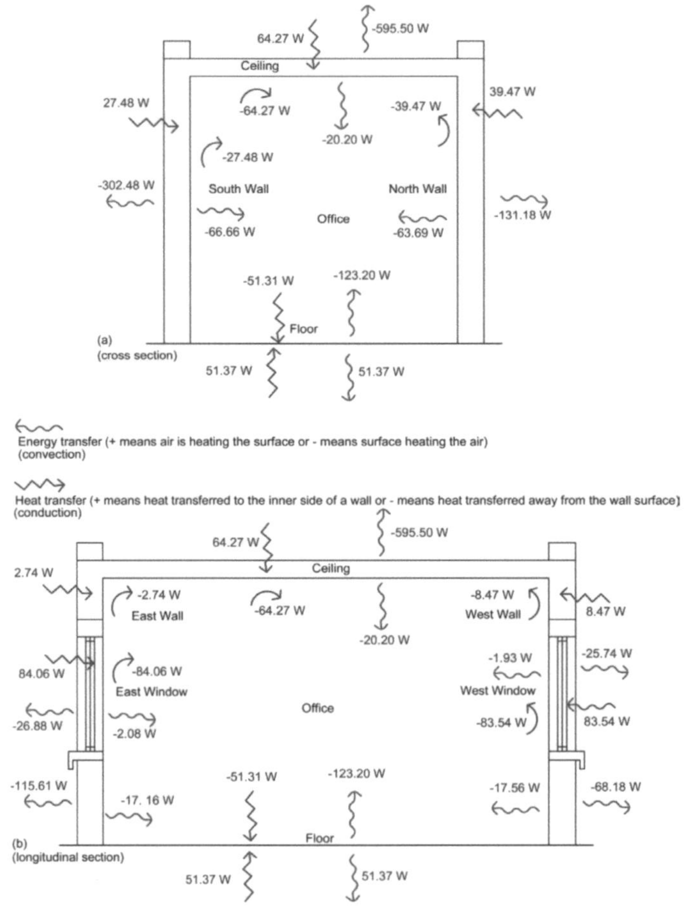

Figure 5.
Yearly average heat and energy transfer performance (for all window opening percentages with all window sizes) of the simulated office.

Plastered solid brick wall	External/ internal solar reflectance	External/ internal emissivity	Conductivity (W/m °C)	Convection coefficient (W/m^2 °C)	Vapor diffusion factor	Density (kg/m^3)	Specific heat (J/kg/°C)
12-mm plaster (inside)	0.600	0.900	0.3	0.0	11.000	960.0	837.0
200-mm solid brick wall	0.350	0.900	0.317	0.0	14.800	1040.0	1050.0
12-mm cement-based plaster	0.600	0.900	0.3	0.0	11.000	960.0	837.0

Flow direction	Internal R value (m^2·°C/W)	External R value (m^2·°C/W)	Internal U value (W/m^2·°C)	External U value (W/m^2·°C)
Horizontal	0.971	0.881	1.03	1.135
Upward	0.911	0.851	1.098	1.175
Downward	1.051	0.921	0.952	1.086

Table 1.
Solid wall properties of the case study building.

DOI: http://dx.doi.org/10.5772/intechopen.101077

Clear 6-12-6 double glazing low E	Dimensionless parameters							Convection coefficient ($W/m^2 °C$)	Specific heat (J/kg·°C)
	Solar transmittance	External solar reflectance	Internal solar reflectance	External emissivity	Internal emissivity	Conductivity (W/m·°C)	Vapor diffusion factor		
6-mm clear glass	0.630	0.200	0.150	0.120	0.845	1.0	99,999.000	0.0	0.0
12-mm air	0.000	0.000	0.000	0.000	0.000	0.0	1.000	2.08	0.0
6-mm clear glass	0.780	0.070	0.070	0.845	0.845	1.0	99,999.000	0.0	0.0
R value (m^2·°C/W): 0.555							U value (W/m^2·°C): 1.803		

Light		Solar energy (EN410)				Pilkington shading coefficients		
Transmittance	Reflectance	Direct transmittance	Direct reflectance	Direct absorptance	Total transmittance (G value)	Short wavelength	Long wavelength	Total
0.760	0.120	0.498	0.193	0.308	0.616	0.573	0.136	0.709

Table 2.
Glass properties of the case study building.

<table>
<tr><th rowspan="2">Months</th><th rowspan="2" colspan="2">Remarks</th><th colspan="2">Yearly averages</th><th colspan="2">Thermally comfortable months (ranges)</th><th rowspan="2">Never thermally comfortable months</th></tr>
<tr><th>External temperature (°C)</th><th>Resultant temperature (°C)</th><th>90% acceptable months</th><th>80% acceptable months</th></tr>
<tr><td>January</td><td rowspan="4" colspan="2">Cool period (below 80–90% ranges)</td><td>10.93</td><td>12.28</td><td rowspan="4" colspan="2">×</td><td rowspan="4">√</td></tr>
<tr><td>February</td><td>12.77</td><td>14.27</td></tr>
<tr><td>March</td><td>14</td><td>16</td></tr>
<tr><td>April</td><td>16.23</td><td>18.51</td></tr>
<tr><td>May</td><td rowspan="6">Warm period</td><td>Within 80–90% ranges</td><td>21.36</td><td>23.86</td><td colspan="2">√</td><td rowspan="2">×</td></tr>
<tr><td>June</td><td>Only in 80% range</td><td>26.03</td><td>28.98</td><td>×</td><td>√</td></tr>
<tr><td>July</td><td>Hot (above 80–90% ranges)</td><td>28.35</td><td>31.21</td><td colspan="2">×</td><td>√</td></tr>
<tr><td>August</td><td>Only in 80% range</td><td>28.43</td><td>30.85</td><td>×</td><td>√</td><td rowspan="3">×</td></tr>
<tr><td>September</td><td rowspan="2">Within 80–90% ranges</td><td>25.70</td><td>27.81</td><td rowspan="2" colspan="2">√</td></tr>
<tr><td>October</td><td>22.83</td><td>24.60</td></tr>
<tr><td>November</td><td rowspan="2" colspan="2">Cool period (below 80–90% ranges)</td><td>17.91</td><td>19.11</td><td rowspan="2" colspan="2">×</td><td rowspan="2">√</td></tr>
<tr><td>December</td><td>13.65</td><td>14.86</td></tr>
<tr><td colspan="2">Averages of the whole year</td><td>Within 80–90% ranges</td><td>19.89</td><td>21.91</td><td colspan="2">√</td><td>×</td></tr>
</table>

Table 3.
The acceptable, cool, and hot months for the simulated office.

Bouchahm [29] identify the optimum window-to-wall ratio (WWR) as 40–50% for energy saving in the Mediterranean climate. Goia [30] found that a WWR between 30–40% is needed for energy saving. Moreover, in this article, the optimum window-to-wall ratio (WWR) is 40% with a 20% window opening for heat loss, thus confirming harmony between the results (10% up or down for different studies), as shown in **Table 2**.

3.3 Findings and discussions

The resultant temperature (operative temperature-°C) is analyzed using ASHRAE 55-2017 [28] to determine when the naturally ventilated office is within acceptable ranges. The averages for May, September, October, and the yearly average are in the range of 80% and 90% acceptability. In addition to these, June and August are only in the 80% acceptability range. Therefore, January, February, March, April, November, and December are considered cool months by virtue of being below the acceptable ranges; but only July is considered hot because it is above the acceptable ranges, as can be seen in **Table 3**.

The yearly average solar gain (for all glass sizes and window opening percentages) of the office is 674.30 W, which also occurs when half of the skin is made of glass with a fully open window. When the glass size on the external skin is changed from 10 to 20%, solar gain increased by 113%. When the glass size on the external skin changed from 20 to 30%, solar gain experienced a 48% increase. When the glass size on the external skin changed from 30 to 40%, solar gain increased by 25%. When the glass size on the external skin changed from 40 to 50%, solar gain increased by 21%. When the glass size on the external skin changed from 50 to 60%, solar gain increased by 20%. When the glass size on the external skin changed from 60 to 70%, solar gain increased by 12%. When the glass size on the external skin changed from 70 to 80%, solar gain increased by 13%. When the glass size on the external skin changed from 80 to 90%, solar gain experienced an 8% increase. When the glass size on the external skin changed from 90 to 100% (full opening), solar gain increased by 8.5%. When the glass size on the external skin changed from 10 to 100%, there was a 763.16% increase in solar gain. Conversely, window opening percentages never affected solar gain, as shown in **Table 4**.

The yearly average infiltration ventilation gain (heat gained or lost by air flow) in the office for all glass sizes and window opening percentages reported a heat loss of 311.49 watts. Regardless of the glass size on the external skin or window opening percentage, the office is always losing heat during hot and humid climatic conditions. When the window opening percentage was set between 10 and 100% (from smallest to largest), the office lost 100% heat when the glass size on the external skin was 10%, 43.24% when the glass size on the skin was 20%, 37.80% when the glass size on the skin was 30%, 30.58% when the glass size on the skin was 40%, 33.78% when the glass size on the skin was 50%, 23.67% when the glass size on the skin was 60%, 28% when the glass size on the skin was 70%, 27.35% when the glass size on the skin was 80%, 38.10% when the glass size on the skin was 90%, and 40.84% when the glass size on the skin was 100%. A maximum of 100% heat loss occurred when the glass size was only 10% with all ratios of window openings, while the minimum of 23.67% heat loss occurred when the glass size was 60% of the external skin with all window opening ratios. However, when the window was 10% opened with all sizes of glass on the external facade (from smallest to largest), the office lost 762% of its heat; when the window was 20% opened with all sizes of glass on the facade, the office lost 622% heat; when the window was 30% opened with all sizes of glass on the facade, the office lost 506% heat; when the window was 40%, 70% and 80% opened with all sizes of glass on the facade, the office lost ~485% heat; when the window was half open with all sizes of glass on the façade, the office lost 458% of its heat; when the window was 60% open with all sizes of glass on the facade, the office lost 473% heat; when the window was 90% open with all sizes of glass on the façade, the office lost 516% heat; and when the window was fully open with all sizes of glass on the facade, the office lost 507% of its heat. The maximum 762% heat loss occurred when the window was always 10% opened with all sizes of glass on the external skin and the minimum 458% heat loss occurred when the window was always opened halfway with all sizes of glass on the external skin, as shown in **Table 4**.

In the hot and humid climatic conditions of Famagusta, the yearly average external environmental temperature of 19.89°C is close to the monthly average of 17.91°C in November. The yearly average resultant temperature (for all external glass skin sizes and window opening percentages) for the simulated office was 21.91°C, which is also very closely achieved when the glass skin is 40% of the opaque skin with a 20% window opening ratio, half of the skin is glass with 40% and 50% window opening

Parameters		Performances	Yearly averages (with all window sizes and openings)
Solar gain	Min.	75.19 W (December-Min.) + 133.04 W (yearly Avrg.) 10–90% window opening with 10% glass on skin	674.30 W
	Max.	1721.18 W (June-Max.) + 1148.66 W (yearly Avrg.) 10–20%, 40%, 60–100% window opening with fully glazed skin	
Heat loss (W)	Min.	−950.23 W (June-Min.) + −607.70 W (yearly Avrg.) Fully opened window with fully glazed skin	−311.49 W (−238 W for optimum WWR because of optimum heat loss when glass size is 40% and window opening is 20%)
	Max.	−24.51 W (January-Max.) + −50.06 W (yearly Avrg.) 10% window opening with 10% glass on skin	
For yearly-based energy reduction (energy saving) optimum WWR is 40–50% according to Badeche and Bouchahm [29]. Moreover, in this article, the optimum WWR is 40% with 20% window opening for heat loss (energy saving).			
Resultant temp. (°C)	Min.	11.47°C (January-Min.) + 20.73°C (yearly Avrg.) Fully open window with 10% glass on skin	21.91°C
	Max.	32.61°C (July-Max.) + 22.94°C (yearly Avrg.) 10% open window with fully glazed skin	
RH for office (%)	Min.	59.02% (June-Min.) + 64.45% (yearly Avrg.) 10% open window with 10% glass on skin	67.56%
	Max.	77.42% (February-Max.) + 68.63% (yearly Avrg.) Fully opened window with fully glazed skin	
Ext. temp. (°C)	Min.	10.93°C (Min.) January (Avrg.)	19.89°C Whole year (Avrg.)
	Max.	28.43°C (Max.) August (Avrg.)	

Table 4.
Performance of the office space in terms of solar gain, heat loss-gain, resultant temperature, relative humidity, and external temperature.

percentages, and 60% of the skin is glass with 80–100% window opening percentages. When the windows are between 10 and 100% opened (increasing the opening ratio), the resultant temperature decreased by 2.58% when the skin is 10% glass, 3.09% when the skin has a 20, 30, 90, and 100% glass facade, 2.75% when the skin has 40% glass, 2.46% when the external skin is half constructed of glass, 3.14% when the skin has 60% glass, and 3.24% when the external skin is 70 and 80% glass on the facade, as shown in **Table 4**. However, when the window was 10% and 80% open with all sizes of glass on the external skin (from smallest to largest), the resultant temperature increased by 7.8%. When the window was 20% open, the resultant temperature increased by 7.63% with all sizes of glass on the external skin (from smallest to largest); when the window was 30% open with all sizes of glass on the external skin (from smallest to largest), the resultant temperature increased by 7.59%; when the window was 40% open with all sizes of glass on the external skin (from smallest to largest), the resultant temperature increased by 8.04%; when the window was 50% open with all sizes of glass on the external skin (from smallest to largest), the resultant temperature increased by 7.73%; when the window was 60% open with all sizes of glass on the external skin (from smallest to largest), the resultant temperature increased by 8.08%; when the window was 70% open with all sizes of glass on the external skin (from smallest to largest), the resultant temperature increased by 7.94%; when the window was 90% open with all sizes of glass on the external skin (starts from smallest to largest), the resultant temperature increased by 7.63%; and when the window was 100% open with all sizes of glass on the external skin (from smallest to largest), the resultant temperature increased by 7.28%, as shown in **Table 4**.

The yearly average relative humidity of the office is 67.56% for all glass sizes on the external skin with all window opening ratios. Moreover, the yearly average relative humidity (aforesaid) is also observed in the office when the glass skin is 10% of the external skin with a 60% to fully open window, when the glass skin is 20% of the external skin with a 40–80% open window, when the glass skin is 30% of the skin with a half open window, when the glass is 40% of the skin with a 20–40% open window, when half of the skin is glass with a 40% open window, when glass is 60–80% of the external skin with a 30–40% open window, and when the glass is 90%and 100% (fully glazed external skin) with a 30% open window. When the window is opened at all ratios (10% to fully opened, starting from smallest to largest), relative humidity increased by 5% when the external skin is 10, 50, and 60% glass, 5.74% when the external skin is 20% glass, 5.96% when the external skin is 30% glass, 5.39% when the external skin is 40% glass, 4.77% when the external skin is 70% glass, 4.52% when the external skin is 80% glass, 4.3% when the external skin is 90% glass, and 4% when the external skin is fully constructed of glass. However, when the external skin has 10–100% glass on the facade (from smallest to largest), relative humidity increases by 2.31% when the window is 10% open, 2.84% when the window is 20% open, 2.5% when the window is 30% open, 2.18% when the window is 40% open, 2% when the window is half open, 1.8% when the window is 60% open, 1.65% when the window is 70% open, 1.5% when the window is 80% open, 1.69% when the window is 90% open, and 1.34% when the window is fully open, as shown in **Table 4**.

Heat is always being transferred away from the internal surface of the opaque wall of the simulated office in all seasons. The minimum heat transfer from the internal surface of the opaque wall for all glass sizes on the external skin and all window opening percentages is in June, during the summer period, at −48.64 W, while the maximum heat transfer is during the winter season in December at −21.26 W. The heat of an opaque wall is always being transferred from the outside surface toward the inside surface in all seasons. The minimum heat transfer from the outside surface of the opaque wall to its inside surface is during the winter season in February at 16.96 W, while the maximum heat transfer is during the summer season in September at 41.31 W. Heat is always being transferred away from the internal surface of the glass wall in the simulated office during all seasons. The minimum heat transfer from the internal surface of a glass surface occurred during the summer season in June at −116.40 W, with the maximum heat transfer during the winter season in November at −58.17 W. The heat on a glass surface is always being transferred from the outside surface toward the inside surface of the glass during all seasons. The minimum heat transfer from the outside surface of the glass to the inside surface is during the winter season in November at 58.17 W, while the maximum heat transfer from the outside to the inside surface of the glass surface is during the summer season in June at 116.40 W, as shown in **Table 5** and **Figure 5**.

The internal surface of an opaque wall in the simulated office always transfers energy to heat the office air during all seasons. The minimum energy transfer from the internal surface of the opaque wall into the office air is in June at −79.07 W during the summer season, while the maximum energy transfer from the internal surface of the opaque wall into the office air is during the winter season in November at −29.12 W. The external opaque wall is always transferring energy to the external air during all seasons. The minimum energy transfer from the external opaque wall into the external air is during the summer season in June at −313.09 W, while the maximum energy transfer from the external opaque wall into the external air is during the winter season in November at −89.89 W. The internal glass surface in the simulated office is always transferring energy to heat the office air during all seasons, except in January, November, and December because energy transfer in the these three months flows from the external air to the internal surface of the glass to heat it. The minimum energy transfer from the internal surface of the glass surface is during the summer season in July at −5.21 W, while the maximum energy transfer from the internal surface of the glass into the office to heat the office air is in October during the summer season at −1.33 W; however, the internal surface of the glass is heated by the office air in January at 1.86 W, in November at 0.69 W, and in December at 0.77, all during the winter season. The external surface of a glass in the simulated office is always transferring energy to heat the office during all seasons, except in January, November, and

		All window sizes with all window openings (10–100%)							
		Heat transfer (conduction)				**Energy transfer (convection)**			
		Opaque (W)		**Glass (W)**		**Opaque (W)**		**Glass (W)**	
		Internal	**External**	**Internal**	**External**	**Internal**	**External**	**Internal**	**External**
Winter season	January	−23.53	27.34	−65	65	−31.19	−94.42	*1.86*	*12.40*
	February	−24.19	**16.96**	−62.05	62.05	−37.41	−160.27	−1.58	−17.38
	March	−30.69	29.25	−84.96	84.96	−52.05	−179.32	−1.74	−20.76
	April	−36.44	29.89	−93.31	93.31	−58.83	−212.97	−2.57	−33.67
Summer season	May	−37.81	27.64	−94.60	94.60	−63.12	−251.98	−4.33	−50.91
	June	**−48.64**	39.69	**−116.40**	*116.40*	**−79.07**	**−313.09**	−4.95	−62.66
	July	−42.08	40.32	−107.46	107.46	−74.97	−307.81	**−5.21**	**−62.92**
	August	−36.42	38.71	−91.38	91.38	−58.59	−253.09	−3.51	−48.36
	September	−35.41	*41.31*	−91.41	91.45	−54.60	−194.54	−2.10	−28.36
	October	−29.28	37.62	−77.47	77.47	−44.76	−167.09	*−1.33*	−15.37
Winter season	November	−21.40	28.33	*−58.17*	**58.17**	*−29.12*	*−89.89*	0.69	6.20
	December	*−21.26*	28.34	−62.02	62.02	−32.06	−96.16	0.77	8.64

Table 5.
Monthly heat and energy transfer performance of the simulated office.

December because in these three months, the energy transfer is from the external air to the external surface of the glass in order to heat the external glass surface using the external air temperature, as shown in **Table 5** and **Figure 5**.

The yearly average heat and energy transfer performances are individually shown including as bold for minimum performances in **Figure 5** for the East, West, South, and North walls.

4. Conclusion

The yearly average solar gain is 674.30 W when the glass size is from 10 to 100% (full) on the external wall with different window opening percentages, although solar gain increased by 763.16% when the glass size on the external skin was suddenly changed from 10% to full (100%) glass. Moreover, the maximum solar gain was observed as 113% when the glass size on the external wall changed from 10 to 20%, while the minimum solar gain was 8% when the glass size on the external wall changed from 80 to 90%.

Office environments with the smallest to largest glass size or the smallest to largest window opening percentage always lose heat; moreover, a 311.49 W heat loss was observed as the yearly average for the above conditions. In addition to this, the window opening percentage never affects the solar gain, as shown in **Table 3**.

The maximum heat loss of 100% occurred when the glass size is only 10% for all window opening ratios, while the minimum 23.67% heat loss occurred when the glass size is 60% of the external skin for all window opening ratios. However, the maximum heat loss of 762% occurred when the window is always 10% opened for all glass sizes on the external skin, and the minimum 458% heat loss occurred when the window is always half opened for all glass sizes on the external skin.

The yearly average resultant temperature for the simulated office is 21.91°C for all glass sizes on the external skin, with all window opening percentages. Furthermore, the yearly average resultant temperature is also achieved when the external skin has 40% glass with a 20% window opening ratio, half of the external skin is constructed of glass with 40 and 50% window opening ratios, and the external skin has 60% glass with an 80% to full window opening ratio.

The yearly average relative humidity of the simulated office is 67.56% for all glass sizes on the external skin with all window opening percentages. In addition to this, the yearly average relative humidity is also achieved when the external skin has 10% glass with a 60% to full opening, the external skin has 20% glass with a 40–80% window opening ratio, the external skin has 30% glass with a 50% window opening ratio, the external skin has 40% glass with a 20–40% window opening ratio, half of the external skin is constructed of glass with a 40% window opening, the external skin has 60–80% glass with a 30–40% window opening, and the external skin is 90% to full glass with a 30% window opening ratio.

During all seasons, heat is always transferred away from the opaque and transparent internal surfaces of the simulated office. Moreover, heat is also transferred from the outside of the opaque walls and transparent surfaces toward both internal surfaces.

Office air is heated by energy transferred from the internal surfaces of the simulated office's opaque and transparent walls during all seasons of the year. The external opaque wall is always transferring energy toward the external air, while the internal surfaces of the transparent surfaces transfer energy to heat up the office air except in

January, November, and December, because the energy transfer in these three months flows from the external air to the internal surface.

Architects, users, and engineers should be careful because in a hot and humid climate like Famagusta, adaptive thermal comfort within buildings is of great importance. Furthermore, July was found to be extremely hot, while January, February, March, April, November, and December were extremely cold, indicating that building users should pay attention to the cost of utilizing mechanical devices in these times of the year.

Conflict of interest

The author declares no conflict of interest.

Author details

Halil Zafer Alibaba
Department of Architecture, Faculty of Architecture, Eastern Mediterranean University, Famagusta, North Cyprus, Turkey

*Address all correspondence to: halil.alibaba@emu.edu.tr

References

[1] Sacht H, Lukiantchuki MA. Windows size and the performance of natural ventilation. Procedia Engineering. 2017;**196**:972-979. DOI: 10.1016/j.proeng.2017.08.038

[2] Wang H, (Yan) Chen Q. Modeling of the impact of different window types on single-sided natural ventilation. Energy Procedia. 2015;**78**:1549-1555. DOI: 10.1016/j.egypro.2015.11.201

[3] Hamdani M, Bekkouche SMA, Benouaz T, Belarbi R, Cherier MK. The study natural ventilation by using buildings windows: Case study in a hot dry climate, Ghardaïa, Algeria. Energy Procedia. 2017;**139**:475-480. DOI: 10.1016/j.egypro.2017.11.240

[4] Yang X, Zhong K, Kang Y, Tao T. Numerical investigation on the airflow characteristics and thermal comfort in buoyancy-driven natural ventilation rooms. Energy and Buildings. 2015;**109**:255-266. DOI: 10.1016/j.enbuild.2015.09.071

[5] Wang J, Wang S, Zhang T, Battaglia F. Assessment of single-sided natural ventilation driven by buoyancy forces through variable window configurations. Energy and Buildings. 2017;**139**:762-779. DOI: 10.1016/j.enbuild.2017.01.070

[6] Alfano FR'A, Olesen BW, Palella BI, Riccio G. Thermal comfort: Design and assessment for energy saving. Energy and Buildings. 2014;**81**:326-336. DOI: 10.1016/j.enbuild.2014.06.033

[7] Frontczak M, Wargocki P. Literature survey on how different factors influence human comfort in indoor environments. Building and Environment. 2011;**46**:922-937. DOI: 10.1016/j.buildenv.2010.10.021

[8] Kisilewicz T. Passive control of indoor climate conditions in low energy buildings. Energy Procedia. 2015;**78**:49-54. DOI: 10.1016/j.egypro.2015.11.113

[9] Halawa E, van Hoof J. The adaptive approach to thermal comfort: A critical overview. Energy and Buildings. 2012;**51**:101-110. DOI: 10.1016/j.enbuild.2012.04.011

[10] Toe DHC, Kubota T. Development of an adaptive thermal comfort equation for naturally ventilated buildings in hot-humid climates using ASHRAE RP-884 database. Frontiers of Architectural Research. 2013;**2**:278-291. DOI: 10.1016/j.foar.2013.06.003

[11] Jin L, Zhang Y, Zhang Z. Human responses to high humidity in elevated temperatures for people in hot-humid climates. Building and Environment. 2017;**114**:257-266. DOI: 10.1016/j.buildenv.2016.12.028

[12] Nguyen AT, Singh MK, Reiter S. An adaptive thermal comfort model for hot humid South-East Asia. Building and Environment. 2012;**56**:291-300. DOI: 10.1016/j.buildenv.2012.03.021

[13] Vellei M, Herrera M, Fosas D, Natarajan S. The influence of relative humidity on adaptive thermal comfort. Building and Environment. 2017;**124**:171-185. DOI: 10.1016/j.buildenv.2017.08.005

[14] TS 825. Thermal insulation requirements for buildings. 2013

[15] Pekdogan T, Basaran T. Thermal performance of different exterior wall structures based on wall orientation. Applied Thermal Engineering. 2017;**112**:15-24. DOI: 10.1016/j.applthermaleng.2016.10.068

[16] Yang W, Zhu X, Liu J. Annual experimental research on convective heat transfer coefficient of exterior surface of building external wall. Energy and Buildings. 2017;**155**:207-214. DOI: 10.1016/j.enbuild.2017.08.075

[17] Li B, Tan M, Liu H, Ma X, Zhang W. Occupant's perception and preference of thermal environment in free-running buildings in China. Indoor and Built Environment. 2010;**19**:405-412. DOI: 10.1177/1420326X10377545

[18] López-Pérez LA, Flores-Prieto JJ, Ríos-Rojas C. Adaptive thermal comfort model for educational buildings in a hot-humid climate. Building and Environment. 2019;**150**:181-194. DOI: 10.1016/j.buildenv.2018.12.011

[19] Zhang Y, Wang J, Chen H, Zhang J, Meng Q. Thermal comfort in naturally ventilated buildings in hot-humid area of China. Building and Environment. 2010;**45**:2562-2570. DOI: 10.1016/j.buildenv.2010.05.024

[20] Indraganti M, Ooka R, Rijal HB, Brager GS. Adaptive model of thermal comfort for offices in hot and humid climates of India. Building and Environment. 2014;**74**:39-53. DOI: 10.1016/j.buildenv.2014.01.002

[21] Cândido C, de Dear R, Lamberts R. Combined thermal acceptability and air movement assessments in a hot humid climate. Building and Environment. 2011;**46**:379-385. DOI: 10.1016/j.buildenv.2010.07.032

[22] Song Y, Sun Y, Luo S, Hou J, Kim J, Parkinson T, et al. Indoor environment and adaptive thermal comfort models in residential buildings in Tianjin, China. Procedia Engineering. 2017;**205**:1627-1634. DOI: 10.1016/j.proeng.2017.10.310

[23] Kumar S, Singh MK, Loftness V, Mathur J, Mathur S. Thermal comfort assessment and characteristics of occupant's behavior in naturally ventilated buildings in composite climate of India. Energy for Sustainable Development. 2016;**33**:108-121. DOI: 10.1016/j.esd.2016.06.002

[24] Barbadilla-Martín E, Lissén JMS, Martín JG, Aparicio-Ruiz P, Brotas L. Field study on adaptive thermal comfort in mixed mode office buildings in southwestern area of Spain. Building and Environment. 2017;**123**:163-175. DOI: 10.1016/j.buildenv.2017.06.042

[25] Singh MK, Ooka R, Rijal HB, Takasu M. Adaptive thermal comfort in the offices of North-East India in autumn season. Building and Environment. 2017;**124**:14-30. DOI: 10.1016/j.buildenv.2017.07.037

[26] EDSL Tas. Software Package for the Thermal Analysis of Buildings. Available from: http://www.edsl.net/main/support/documentation.aspx [Accessed: 10 April 2021]

[27] Location of Famagusta on a Map [Internet]. 2021. Available from: https://www.weather-forecast.com/locations/Famagusta/forecasts/latest [Accessed: 01 October 2021]

[28] The American Society of Heating, Refrigerating and Air-Conditioning Engineers (ASHRAE). ANSI/ASHRAE Standard 55-2017. In: ASHRAE Standard-Thermal Environmental Conditions for Human Occupancy, American Society of Heating, Refrigeration and Air-Conditioning Engineers. Atlanta, GA, USA: ASHRAE; 2017

[29] Badeche M, Bouchahm Y. Design optimization criteria for windows providing low energy demand in office buildings in Algeria. Environmental and Sustainability Indicators. 2020;**6**:1-10, 100024. DOI: 10.1016/j.indic.2020.100024

[30] Goia F. Search for the optimal window-to-wall ratio in office buildings in different European climates and the implications on total energy saving potential. Solar Energy. 2016;**132**:467-492. DOI: 10.1016/j.solener.2016.03.031

Chapter 10

Holistic and Affordable Approach to Supporting the Sustainability of Family Houses in Cold Climates by Using Many Vacuum-Tube Solar Collectors and Small Water Tank to Provide the Sanitary Hot Water, Space Heating, Greenhouse, and Swimming Poole Heating Demands

Luis E. Juanicó

Abstract

This work presents a new proposal for supporting the sustainability of a single-family house in very cold climates by installing many vacuum-tube solar collectors and a small water tank in order to fulfill the whole dweller demands of heat: space heating, sanitary hot water, and warming both, a greenhouse (spring and autumn) and a swimming pool (summer). This way is obtained a sustained demand that maximizes the utilization of heat from solar collectors throughout the year. This system is designed intending to use the smallest tank that fulfills the winter heating demand, supported by vacuum-tube solar collectors and a little help from electrical heaters working just on the valley tariff. This innovative design gets the most sustainable (but affordable) solution. This goal can be achieved by using a small well-insulated overheated aboveground water tank, instead of the huge underground reservoir of heat used by most projects tested up today. These large communal projects use huge reservoirs to provide seasonal thermal storage (STES) capacity, but their costs are huge too. Besides, it was observed that all these huge STES suffer large heat losses (about 40%), due to constraints for thermally insulating such very heavy systems. On the contrary, our small aboveground water tank can be thermally insulated very well and gets affordable costs. In this work is developed dynamical solar-thermal modeling for studying this novel approach and are discussed its major differences with traditional design. This modeling is used to study the whole demands of heat for one family living in the same conditions of the Okotoks' project. The Okotoks' project is based on many flat solar collectors (2,290 m^2) and a huge (2,800 m^3) rocky-underground STES system in order to almost fulfill (97%) the space heating demand of 52 houses (15,795 kWh/y ea.) in Alberta (Canada), having an overall cost of 9 MU$ (173,000 U$ ea.). We have already shown in previous work that this new proposal could reach noticeably lower costs (€30,500) than the Okotoks' project in order to provide the same heating demand, by taking advantage of using 18 vacuum-tube collectors (solar area 37 m^2) and a small (72 m^3) well-insulated (heat losses 18%) water tank heated up to 85°C, which is the same temperature used in Okotoks and other traditional projects. Now, this proposal is enhanced by using a holistic approach to include other low-temperature demands (sanitary hot water and warming a greenhouse and swimming pool) that enhance the sustainability of dweller living. This way, the full production of heat from solar collectors is utilized (about six times larger than the single space heating demand, but using only 20 vacuum-tube solar collectors (21 m^2 solar area) and a very small (10m^3) water tank, reaching about a lower overall cost (€20,000), and so, the economic performance is enhanced as well. Besides, it is shown that using a small fraction of electrical heaters as a backup system (2%) and slightly overheating the water (up to 120°C@2 bar), which is feasible by using commercial stainless steel water tanks designed for such purposes, its economic performance could be again noticeably enhanced (reducing the overall cost to €20,000, and getting payback period less than two years). This way here is demonstrated the overall solar-STES system can be reduced by about half size meanwhile the energy output can be increased up to seven times. Hence, the thermal analysis performed suggested us strongly critic the traditional approach of using flat solar collectors instead of vacuum-tube collectors. This analysis shows that this choice has strongly driven the selection of a huge STES, which in turn increases noticeably the overall costs of the system since for such huge STES is mandatory to use underground reservoirs. However, this analysis also shows that without including those secondary demands, this proposal achieves a modest economic performance (payback period about 11 years) regarding its lower energy saved and compared against the "most smart" standard solution (one water tank with electrical heaters, costing about 5,000 U$ and exploiting the valley tariff of nocturnal electricity costing 0.1 €/kWh). On the contrary, when these secondary demands are included, the payback period is reduced by two years. Beyond the particular case studied here, this analysis suggests that the right design of any solar + STES system should be led by the solar production. On the contrary, the traditional design intends to fulfill one demand (space heating) concentrated during winter, and so, its performance is noticeably penalized, and the solution is definitely not to put a larger tank. Unfortunately, up today the poor performance of these projects has shown that this solar technology is (by far) unaffordable. Maybe its best days have gone, considering the enormous improvements achieved by another solar technology (using photovoltaic panels + heat pump + small daily-storage water tank), as it was discussed here.

Keywords: zero-energy buildings, seasonal thermal energy systems, solar collectors, household energy demand, thermal and economic analysis

1. Introduction

1.1 Previous works

In a previous work [1], we have already developed a dynamical thermal modeling of solar collectors linked to a seasonal thermal storage (STES) system. The major advantage of this model is its simplicity. This model uses one-step time and explicit numerical scheme so that it can be programmed easily on any standard spreadsheet, instead of other complex three-dimensional codes (like TRANSYS). Besides, in this work was deeply discussed that a complex code is completely unnecessary for modeling STES systems based on well-insulated water tanks. On the contrary, it was demonstrated by using physical simplifications that such systems can be modeled as a zero-dimensional system, and so, this simplest (the so-called lumped-capacity model) spatial modeling can be used, in which all the water tanks can be considered at the same homogeneous temperature. Furthermore, this work has also demonstrated that a very refined time simulation neither is necessary, since a large (seasonal) storage system would have only slow variations of temperature (cooled during winter and heated during summer) and so, the STES yearly evolution can be perfectly modeled by using one-day time step. Beyond these useful simplifications, the major strength of this simple model is to provide a useful framework for modeling the solar system (a set of solar collectors) together with the STES system, which in turn allows us to take-in-hand all the system parameters. These parameters comprise the solar collectors' parameters (their number, title angle, and their efficiency's equation), as well as the STES parameters (the water storage capacity, the insulation quality, and maximum/minimum working temperatures). Hence, the analysis performed has surprisingly shown that the behaviors of the STES and solar systems are actually coupled. This way, we have found that a small short-term STES working with many vacuum-tube solar collectors can provide the same overall performance (that is, to fulfill the space heating demand during winter) that a huge seasonal STES working with many flat solar collectors, meanwhile the first design reaches a noticeably lower cost.

In this work will be summarized the major findings obtained in this previous work [1] as the starting point for the present analysis. On this, it will be studying a novel proposal that could enhance the performance of this novel design. It uses an overheated water tank (up to 120°C) instead of the 85°C level previously used, which creates a slight overpressure (2 bar) that can be withstood by using commercial stainless steel tanks. Besides, this proposal follows a holistic approach in order to include all the secondary heat demands related to a family living in a very cold climate location, besides the space heating demand that is concentrated during winter. These demands comprise the sanitary hot water demand, warming a greenhouse (from spring to autumn), and swimming pool (during summer). Therefore, a sustained demand throughout the whole year is created. This is a key to maximizing the production of energy from solar collectors since the small STES only can provide a short-term (less than one month) storage capacity. This constraint was a serious drawback shown in the previous study, in which the surplus of solar energy produced by solar collectors must be avoided for ten months each year. Hence, we have realized that such kind of solar + STES system could be better utilized since it uses a very large number of solar collectors to fulfill the heating demand concentrated during winter, but its large potential for generating heat during the whole year is not exploited. Here was the starting point for the holistic approach (by including all the secondary demands of heat) studied here.

1.2 Present development of the (solar plus thermal storage) technology

During the last twenty-five years have been developed several testing projects of Seasonal Thermal Energy Store (STES) associated with a solar system based on many collectors for providing space heating to many houses (or department buildings) in cold locations, mostly in Canada and Germany [2, 3]. These projects have shown this technology is feasible, although very costly.

The first and most important project that will be considered here is the well-known Okotoks' project, developed by the Drake Lake Solar Community in this place (51°N) within the Alberta State of Canada. This project intends to create a sustainable solar community, and their performances are available online by internet, as well as by several technical reports. For example, Sibbitt et al. [4] has recorded their (solar and thermal) performances. This project works since 2007 up today, and it provides annually 97% of the space heating demand to 52 well-insulated houses (117 kWh/m^2/y and 135 m^2 ea., in this very cold climate (average 5.2°C and 5,020 heating degree days defined according to [5]). This project uses 798 flat solar collectors (2,290 m^2) and a long-term storage system that heats the underground rock by means of 144 very deep (40 m depth) wells drilled covering over a 700 m^2 field (that is, covering a 28.000 m^3 volume of rock). This huge STES system is designed to provide seasonal storage (that is, the solar collectors accumulate heat during summer, and houses demand heat during winter), but this design (using the rocky underground) only achieves an overall efficiency of 60%, since there are remarkable heat losses from the reservoir (heated up to 90°C) to the surrounding ground. This huge long-term STES system works together with another short-term system based on 240 m^3 water tanks, which provides the household space heating by using under-floor water systems. This is the right choice in order to maximize the thermal capacity of the reservoir (working within the usable 85–35°C range, instead of working within the usable 85–60°C range when hot-water radiators are selected as the heating system). The solar radiation received on the 45°-inclined collectors (13,902 GJ/y) is collected with an overall 31% efficiency, and can effectively store solar energy only during the warm season [6]. The heating demand in Okotoks is very concentrated during four months (94%), which is a typical characteristic of cold continental climates. These figures present an exigent scenario for a STES system that explains the superlative cost (U$173,000 per each house) of this project [7], mostly due to the extremely high cost of constructing the rock reservoir.

In previous work, we have already analyzed our novel approach for solving the heating demand of a single house on the Okotoks′ project. However, since this project uses a different kind of STES system, we need to state another two starting points for performing our study. For our purpose, just let us keep in mind the major parameters for every single house of the Okotoks' project: the cost (U$173,000 ea.), the solar collector area (44m^2), the rocky reservoir (538 m^3), and finally, the annual heating demand (15,795 kWh/y). These parameters will be useful for comparing after with other designs.

The second reference point considered here is the Friedrichshafen (48°N) project, working since 1996 in Germany for heating a department building. This project considered a STES system based on a huge (12,000 m^3 and 20 meters height) underground water tank. This tank is also very heavy since it is built by using 60 cm-thickness reinforced-concrete walls that include a 1.2 mm stainless steel liner. Although this STES system is huge, it can satisfy, only partially (just 25%), the space heating demand of a multifamily building (23,000 m^2, 100 kWh/m^2/y) by using hot-water radiators. The solar system comprises 4,050 m^2 flat solar collectors installed onto 38° inclined roofs [8, 9]. This project has preferred to use a huge water tank in order to reduce its area/volume ratio and so their specific heat losses and cost. This goal was achieved when it is compared against other similar (but smaller) German projects. So, this (12,000 m^3) tank achieves a lower specific cost (112 €/m^3) than the Hamburg project (4,500 m^3,

220 €/m^3) and the Hannover (2,750 m^3, 250 €/m^3) project [2]. However, due to its heavy mass and large depth, it is also very difficult to wrap this tank with standard isolative materials, which can withstand pressure up to 2 bars. So, the Friedrichshafen project has recognized heat losses of about 40% on this huge tank related to the lack of thermal insulation on its lower third (bottom and walls). Also by considering its heat losses of about 8% in the heat distribution system. This project shows the drawbacks of building a huge communal system, regarding our approach that designs a small system for each house. Besides, regarding the use of flat collectors working up to 90°C in cold climates, this project demonstrated that these collectors can achieve a poor average efficiency (30%). The total investment of this project (4 M€) is recognized by Bauer et al. [8] as a not cost-effective solution, regarding the low percentage (25%) of fossil fuel substituted. Let us note that by comparing the heat productions of the Friedrichshafen and Okotoks projects, the German project achieves an equivalent cost of about 128,000 dollars per Okotoks' house (considering an exchange conversion of 1.2 dollars per euro). So, even recognizing that the German solution is cheaper than Okotoks, it is not enough cheap to become an affordable solution by far. However, there are some interesting learned lessons obtained from these German projects; the feasibility of using water tanks as the main thermal storage system (the cost of this huge tank is about 66,000 dollars per each equivalent Okotoks' house), and the worse performance of using hot-water radiators instead of in-floor water as space heating system, regarding the poor yield obtained from flat solar collectors. The main parameters of this STES system can be calculated in order to compare against the Okotoks one, by taking its equivalent heating demand related to a single Okotoks' house (15,795 kWh/y). Hence, we can obtain: an overall cost of 128,000 dollars, STES water volume of 320 m^3, and solar collector area of 108 m^2. These poor numbers reflect the bad choice using a high-temperature heating system (hot water radiator instead of underfloor hot water) and shows the coupling effects between the three (solar, STES, and heating) systems involved.

The third project that we consider now as a reference point uses a small water tank for heating a single house. The Irish Galway project was initiated in 2006, [10]. It uses a 23 m^3 underground water very well insulated (wrapped by an EPS layer of 60 cm thickness) and six vacuum-tube collectors (2 m^2 solar area and costing €500 each one) for heating a single house (1,827 kWh/y) within a temperate climate (2,063 heating degree days). This project is important for us because it has demonstrated the economic feasibility of small solar+STES systems, which can reach reasonable investments (€ 28,344). In addition, from the detailed cost breakdown performed by Colclough and Griffiths [10], it is obtained a good starting point for developing now our economic analysis. For example, this project has shown that large fixed costs (€4,300) related to the many auxiliary systems (temperature sensors, valves, piping, controller, pumps, etc.) required, and the same fixed cost will be considered in our project.

Besides, the Galway's project provides some useful lessons:

1. The actual cost of the underground tank exceeds largely the sole cost of the stainless-steel tank (€ 5,350). The total cost of the water tank must include their insulation (€ 3,060), the soil excavation (€ 1,404), and other labors related to the underground sitting (the construction of a grave and another impervious layer) add € 7,800, increasing the total cost up to € 17,600.

2. Thermal stratification does not occur within this well-insulated tank, in which have been measured temperature differences down 2°C.

3. The falling prices of solar collectors and the relatively high cost of the solar installation (€900), has been recognized by Colclough as a reason for installing more collectors since the installation cost is almost a fixed cost. Following this concept, Colclough, Griffiths, and Smyth [11] have estimated by numerical calculation that the solar fraction of the heating demand could be increased by 50% by doubling the number of solar collectors, which implies a modest extra investment of €3,000 when it is compared against the overall cost.

Regarding the last point from Colclough, we can expect significant improvements by performing an economical optimization on the number of collectors. This analysis should be done by considering the performance of both, the solar system and the STES system. However, at present, there is not any modeling tool available for this purpose. Most of the works have performed thermal models of STES by using complex numerical codes, like TRNSYS or ANSYS [11–13]. However, regarding the high complexity of these tools, we have realized that these codes are not suitable for modeling altogether the solar and thermal behaviors and for taking all-in-hand its parameters, as it is provided in this work by developing an explicit numerical model. Otherwise, the TRNSYS and ANSYS codes are suitable for modeling systems having two main characteristics:

1. Fast-transient dynamic, in which a very-detailed time discretization is required, which can be solved by using a time step of about one minute.

2. Spatial gradients of temperature are relevant, which can be performed by using finite volume method.

The first characteristic is not actually relevant for modeling large-term (as seasonal) STES systems, in which the evolution of the tank temperature is very slow, according to the high ratio between energy stored and energy demanded every day. Therefore, in such kinds of systems is not necessary to consider time steps shorter than a day. The second characteristic is relevant for modeling huge underground STES systems, in which their large weight forbids us to insulate their bottom part as it occurs in the aforementioned Friedrichshafen's project. However, this is not the case with small tanks, as it is proposed here. Those heavy STES systems suffer noticeable heat losses, and high-gradient temperature profiles, in both, radial and axial directions. On the contrary, this behavior can be neglected within small well-insulated tanks, as the Galway project has demonstrated [10].

From these findings, we have developed a simple lumped-capacity thermal model for water tanks, which assumes that both (radial and axial) temperature profiles can be neglected and so, all the water can be considered as having the same (homogeneous) temperature. The axial profile can be minimized by putting the source heat exchange below the sink one (this is the opposite configuration usually used in huge tanks, which is created a stratified temperature profile in order to minimize the heat losses about the not-insulated bottom part of the tank), and so, causing a free-convection flow that counterbalances the stratification, as the Galway's project has conveniently used [11]. In addition for aboveground tanks, there is a uniform boundary condition (the outdoor temperature) that helps to create a homogeneous axial profile. So, the radial temperature profile could be neglected in small tanks; indeed, this effect not solely depends on the tank size. Regarding the very-well known thermal behavior related to the heat conduction within a body surrounded by a fluid convective cooling [14], the diffusion of heat along the radial axis is complemented by the convective heat losses at the outside surface of the tank: In "large" tanks the heat diffusion is relevant and the convective heat transfer can be neglected. Meanwhile, the opposite behavior occurs in "small" tanks; but, indeed, their relative importance (their quotient) is actually represented by the dimensionless Biot number:

$$Bi = h\,D/\lambda_w \tag{1}$$

where h is the convective coefficient of heat transfer (at the external surface of tank), λ_w is the thermal conductivity of water, and D is the tank diameter. In general, problems involving Biot <1 (in which the heat diffusion within the tank can be neglected) are simple, since they can be considered that the temperature field within the tank is homogeneous [14], and so, the single thermal resistance (and temperature variation) to be considered is the one related to the boundary convective layer. Let us note that for well-insulated aboveground tanks, the convective coefficient (h) does not involve solely the thermal resistance of the convective film layer; otherwise, it rather than represents the thermal resistance of the insulation layer (that is, the major thermal resistance involved here), defined by its thermal transmittance (U). The reader can check that for the largest tank considered here (D = 3 m, λ_w = 660 W/K.m, $U = h$ = 0.1 W/m^{2}K), it is verified that Bi <1. Indeed, even observing some minor temperature difference, as the 2°C difference measured in the Galway's tank [10], it must be considered that the actual temperature of the tank's surface is always lower than the mean temperature of the tank, and so, this homogeneous model is conservative for estimating the heat losses. Besides, by placing the sink heat exchanger on the central axis, the heat is delivered to the house with a temperature higher than the average, and so, this model is again conservative.

2. Solar and thermal modeling

2.1 New system design

This conceptual design considers many vacuum-tube solar collectors for heating one water tank up to 120°C in order to provide space heating by water in-floor system. Regarding previous works (up to 85°C), this overheating can be achieved with a modest tank overpressure (2 bar) that can be easily withstand by commercial stainless steel tanks (designed with a relief valve at 3 bar), meanwhile, this tank doubles the useful heat capacity (from 120–33°C) of previous tanks (from 85–33°C). So, the water-glycol mixture is heated up to 125°C and the in-floor system is cooled up to 28°C in order to maximize the working range of temperature within the tank, by considering a 5°C temperature jump in both heat exchangers, similarly to the Galway's project. This 5°C difference is enough for using standard tubular-copper exchangers that provide the demanded (~10 kW) heat power while getting affordable costs [15].

On the other hand, our design intends to use a small tank having a storage capacity of between two to four weeks for the winter heating demand. A smaller tank has a lower cost and also, a smaller total area, which in turn implies lower heat losses and insulation cost. This small tank is designed to be heated only around one month previous to the winter demand in order to be ready for this exigent demand, but most part of the year this tank is actually not used, meanwhile the vacuum-tube solar collectors are used for heating the secondary demands. Let us note that, this kind of collector has a remarkable ability for collecting energy even during cloudy days. For instance, according to measured data of the vacuum-tube collector manufactured by Apricus, its yield during cloudy days is 25% of the yield obtained during clear days [16]. On the other hand, a flat collector would have a negligible yield during cloudy days, and even on sunshine days during cold winters.

2.2 Solar collectors modeling

The use of vacuum-tube collectors in order to maximize the solar yield during winter is a key within this design, instead of the flat collectors usually used in these tested projects. This point will be discussed now by considering the efficiency curve of commercial vacuum tubes and flat solar collectors (see **Figure 1**), which are provided by the European Solar Industry Federation [17]. The instantaneous efficiency (η) of any solar collector can be approximated by its optical efficiency (a_0) and their linear (a_1) and quadratic (a_2) heat-losses coefficients, as is described by Eq. (2). Here, T_m is the mean temperature of collector, which receives normal solar flux, I_n (W/m^2), and T_a is the ambient temperature [18].

$$\eta = a_0 - a_1(T_m - T_a)/I_n - a_2(T_m - T_a)^2/I_n \tag{2}$$

Let us note in Eq. (2) that both heat-losses terms are divided by the normal irradiation (I_n). Hence, regarding that flat collectors have higher heat-losses coefficients than vacuum-tube ones, this effect (penalizing flat collectors) is minimized in **Figure 1** by considering a very high (800 W/m^2) I_n value, for which both curves intersect at 70°C (by taking the gross area for the vacuum-tube collector, which is another subjective decision that clearly favors flatting collectors). However, this value does not represent by far an actual average condition. Although this flux could be observed as the total solar irradiation (I) on clear days, a flat collector would obtain

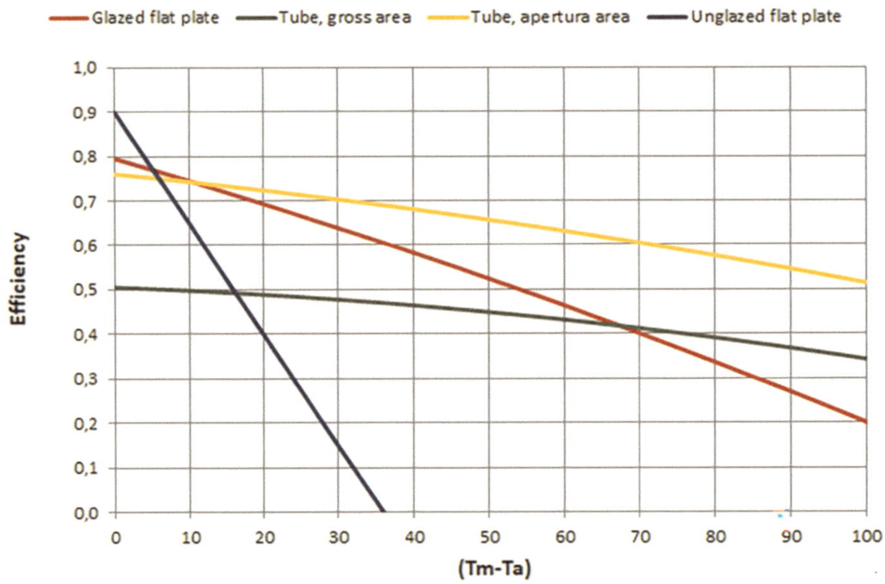

Figure 1.
Efficiency curve for different kinds of solar collectors.

this flux as its normal projection (I_n) only at noon (when its azimuthal angle is null) and twice along the year (when the elevation angle of the sun above the horizons matches normally the collector's tilt angle). Therefore, it is more accurate to consider both efficiency curves for lower I_n values. For instance, **Figure 2** are illustrated the efficiency curve for both collectors working on I_n = 400 W/m^2 and 200 W/m^2, for which are obtained intersecting points of 34°C and 17°C, respectively. These results show that, in these cases, the flat collector almost never gets higher efficiency than the vacuum-tube collector. In addition, for this last case (I_n = 200 W/m^2), the flat collector cannot get energy for temperature differences higher than 37°C, meanwhile, the vacuum-tube collector still gets a remarkable 36% efficiency in this case.

Let us remark that these low values of solar normal flux do not represent necessarily a cloudy-day condition. For example, let us consider now a fully sunny winter day (I = 800 W/m^2) in the Friedrichshafen project (having 38°-inclined collectors) at 4 pm, that is, when the elevation angle of the sun over the horizon is 2° and thus, the zenithal-angle of the sun with the collector's normal is 50°. For this condition, the sun rays present an azimuthal angle of 60° onto a south-oriented flat collector. So, the product of their cosines (cos60° x cos50° = 0.32) leads to getting a normal irradiation flux over the flat collector I_n = 257 W/m^2, for which the maximum temperature difference, this flat collector could reach is 47°C. Therefore, it can be inferred that this flat collector always would obtain negligible winter yields working with 55°C hot-water radiators, as was observed in the Friedrichshafen's project. Otherwise, in this case, the vacuum-tube collector gets 34% efficiency, which is calculated by taking I_n = 514 W/m^2, regarding that for this case, the azimuthal projection (cos60°) must not be considered, due to this cylindrical geometry. This comparison can be extended, for example, to a fully sunny summer day at 4 pm when the elevation angle of the sun is 17° and then, the azimuthal angle over collectors is 35°, leading to I_n = 327 W/m^2 for the flat collector and I_n = 654 W/m^2 for the vacuum-tube collector.

Therefore, following the previous discussion, we can conclude that **Figure 2** induces us to make a huge mistake that is to compare both collectors as working on the same I_n value. Another way of saying this is that the crossing-point of both curves cannot be used at all as a criterion for comparing the performance of flat and

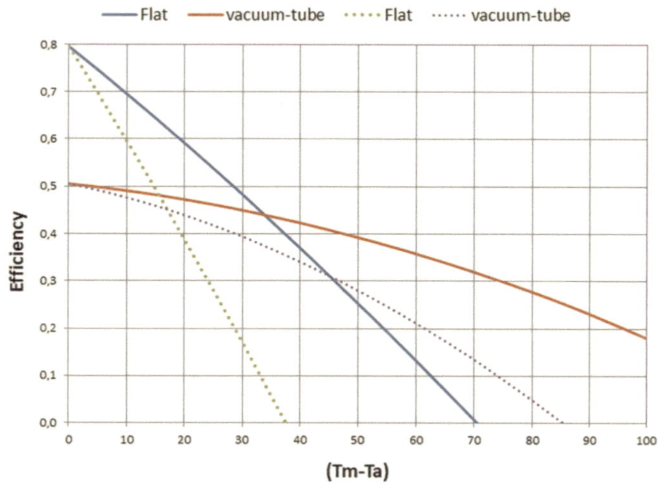

Figure 2.
Efficiency of flat and vacuum-tube collectors (I_n = *400 and 200 W/m*^{2}*).*

vacuum-tube collectors. Let us note that a flat collector receives a variable azimuthally-projected solar area along the day, meanwhile, a cylindrical collector always offers the same azimuthally-projected solar area. Although a full discussion of this issue depends on many factors, such as the day of the year and the latitude of the location, etc., maybe we could consider now a useful analogy. Regarding the total solar energy received along the day (G, kWh/m^2), the flat collector can be represented by a fixed PV panel, and meanwhile, the vacuum-tube collector could be represented by another PV panel mounted onto a one-axis solar-tracking system. Hence, we can compare the yield of both collectors by using the well-known result that the one-axis tracking PV panel produces about 30% more energy than the first fixed PV panel [19]. Therefore, and taking into account that our solar model calculates the G received for tube collectors and not for flat collectors (see the solar_trajectory.xls file in [20]). It will be conservatively estimated the G values received by flat collectors by reducing 25% the G values calculated for cylindrical collectors. And now, let me be completely clear about this point. In my opinion, it is completely unforgettable that most solar researchers have traditionally neglected this mismatch behavior between flat and vacuum-tube collectors; I guess that this is due to some aversion against vacuum-tube solar collectors, which mostly are manufactured in China. However, and in order to be fair for both kinds of collectors, I have to note also that vacuum-tube solar collectors have a major drawback, regarding their concerns about overheating, which potentially can be very dangerous (especially considering heat-pipe collectors with an integrated water tank above vacuum tubes), but, precisely this kind of solution as is studied here (by using large water tank and controlling system) should be the "silver bullet" for this weakness. In my opinion, from the selection of flat solar collectors within most projects performed up today, we can realize that this aversion exists. As we will discuss here, the huge costs reached for all projects performed up today (by using a huge STES system) are related to this choice, and this is the major cause of the failure of this technology. A failure that may cannot be overpass in the future, since nowadays is been coming to another solar technology with a better perspective for solving the heating household demand. This novel technology is the air/water heat pump (having efficiencies around 400%) that can be linked with photovoltaic panels in order to get a sustainable solution, as well as the proposed (solar + STES) technology, does.

2.3 Thermal model of STES

The STES system is modeled on these useful assumptions for simplifying:

1. The condition of temperature surrounding aboveground tanks is modeled by using the monthly averages of outdoor ambient temperature. This assumption is reasonable by considering that the tank temperature varies slowly, and so, any fast variation on the ambient temperature is counterbalanced and can be neglected in order to calculate the monthly heat losses of the tank.

2. The fully time-related terms involved in the STES energy balance (that is, thermal powers and temperatures) are considered by means of their monthly averages. This is a reasonable assumption that the heat transfer mechanisms (conduction and convection heat losses) involved are described by linear equations and so, they both are proportional to the difference of temperatures. Therefore, any fast fluctuations are counterbalanced when this term is integrated along one month. However, let us point out that by considering a numerical

scheme based on monthly time steps, it will be introduced several numerical errors, and for this reason, after performing this model, another model will be performed by using daily time steps in order to verify the accuracy of the results obtained with the previous monthly model.

3. The field of temperatures within the water tank can be considered homogeneous, T. This assumption is reasonable according to its low Biot number, as it was previously discussed.

Working on the previous three hypotheses, the time evolution of the tank temperature can be calculated by means of the thermal model based on lumped capacities [14]. Taking this model and by considering now the energy equation, the rate of internal energy can be calculated by counterbalancing the solar power (Q_{solar}) gained with both extracting powers, the heat losses to the surrounding ambient (Q_{amb}) and the power delivered to the household in-floor heating system (Q_{heat}):

$$M\,c(dT\,(t)/dt) = Q_{solar}(t) – Q_{amb}(t) – Q_{heat}(t) \quad (3)$$

In this equation, the water mass and its heat capacity are noted by M and c, respectively. Thus, the annual evolution of the STES temperature can be described by using their monthly averages going through a twelve-equation system, as:

$$M\,c\;(dT_1/dt) \sim= M\,c\;\left(T_2{}^1 - T_1{}^0\right)/\Delta t = Q_{solar}{}^1 – Q_{amb}{}^1 – Q_{heat}{}^1 \quad (4)$$

$$M\,c\;(dT_2/dt) \sim= M\,c\;\left(T_3{}^1 - T_2{}^1\right)/\Delta t = Q_{solar}{}^2 – Q_{amb}{}^2 – Q_{heat}{}^2 \quad (5)$$

$$M\,c\;(dT_{12}/dt) \sim= M\,c\;\left(T_1{}^1 - T_{12}{}^1\right)/\Delta t = Q_{solar}{}^{12} – Q_{amb}{}^{12} – Q_{heat}{}^{12} \quad (6)$$

In this system is approximated every *n-teenth* (n = 1,2 ... 12) temperature rate (dT_n/dt) by its difference quotient along its monthly time step, $(T_{n+1} - T_n)/\Delta t$, and so, we can transform the original system of differential equations in another (simpler) system based on algebraic equations. The supra-index in powers and the sub-index in temperatures denote the ordinal monthly number, and the supra-index in temperatures denotes the number of the numerical iteration performed for obtaining an accurate solution. This system describes an explicit numerical scheme that can be solved by performing an iterative procedure. Thus, starting with a seed value for the first month ($T_1{}^1$), the first value of every month can be cleared going through (Eqs. (4)–(6)). Here, the last Eq. (6) gives us the new value for the first month ($T_1{}^2$), from which this iterative procedure starts again, and continues until the whole system converges to the stationary periodical solution, which describes the temperature evolution of the tank. This numerical code was performed on a spreadsheet (see Complementary Material, in [20]) and from our results, we have observed that usually, only little iterations are needed. This behavior is expected, according to the physical characteristic of this system, which is an energy dissipater. So, we have chosen a spreadsheet instead of any procedural language for developing our numerical tool, considering its advantage of providing explicit all-in-hand modeling.

We have to calculate now the three power terms, Q_{solar}, Q_{amb}, and Q_{heat}, in order to solve the previous algebraic system (Eqs. (4)–(6)). The first term (solar power) depends on the solar resource and the efficiency of solar collectors. For N collectors having collecting area A_c and instantaneous efficiency $\eta(t)$, the instantaneous solar power collected from a normal solar flux $I_n(t)$ is given by:

$$Q_{solar}(t) = N\,A_c\,\eta(t)\,I_n(t) \tag{7}$$

Actually, we are only interested in obtaining the monthly averages of the collected solar energy, $E_{solar}{}^n$. So, the monthly balances of powers (Eqs. (4)–(6)) will be substituted by the monthly balances of energy. However, it is cumbersome to integrate Eqs. (7), due to the high variability of the solar resource. On the other hand, the single solar data worldwide available are the monthly averages of the daily solar energy on ground level (G^n). Thus, this lack is now solved by introducing monthly average factors (α^n x G^n), which take into account the relation between both monthly solar irradiations, the received on ground level and the received on the collector when it is elevated a given tilt angle (φ). By using these factors, the monthly average of the collected energy can be calculated by:

$$E_{solar}{}^n = N\,A_c\,\eta^n\,\alpha^n\,G^n \tag{8}$$

In this equation, we have been introduced the monthly averages of the collector's efficiency, η^n, which (from Eq. (2)) can be calculated by substituting the actual collector mean temperature (T_m) at the *n-teenth* month by the tank temperature calculated at the previous month, $T_{n\text{-}1}$:

$$\eta^n = a_0 - a_1(T_{n-1} - T_a)/I_n - a_2(T_{n-1} - T_a)^2/I_n \tag{9}$$

Here, let us note that the mean collector's temperature is around 5°C, higher than tank temperature by taking into account the temperature jump in the heat exchange. However, also the ambient temperature could be considered around 5°C above its daily average, according to the fact that the collector gets its highest efficiency mostly around noon, and so, these opposite effects are canceled.

The α^nxG^n factors introduced in Eq. (8) must be calculated according to the latitude of the location, the day of the year and the collector's shape (cylindrical or flat), and they can be calculated from many solar codes available in the literature. Here is provided with a code programmed for cylindrical collectors (see in solar_trajectory_Bariloche.xls [20]), based on the well-known equations that describe the apparent trajectory of the sun [1, 21]. By using this code, the procedure from which the calculation of the α^nxG^n factors can be described by four steps:

1. To calculate the sun trajectory and its normal collector's area along the day for every month. This step is performed by setting the input data (cells B4-B7) by considering the mean day of every month (for example, d = 15 for January, etc.). Thus, cells A14: F255 are obtained the intended results.

2. To calculate for every month the daily solar energy received on the ground (G) for a given (constant) direct solar irradiation, I. This calculation is performed by setting to zero the collector's tilt angle (cell B6) and so, the calculated G value is obtained in cell B10.

3. By using this I-G calculation and the known values of G^n, now we can calculate (by iterating) the monthly values of I, that is, the equivalent direct solar irradiation that provides the same energy on the ground. Of course, this is a simplified model that does not consider the diffuse solar irradiation (that is an

isotropic term), but, for the goal looking for here (to get the monthly averages of collector's production), this is a reasonable approximation.

4. Then, by using these monthly I values it is calculated the solar irradiation that would receive a cylindrical collector (α^nxG^n) mounted with a given tilt angle.

Although at a first glance this procedure could be cumbersome, it is a well-known methodology; there are many similar software applications for calculating the solar irradiance over a collector as a function of its tilt angle. For example, the reader can study the simulating tool developed by NASA [22] for studying the G value for different tilt angles in any worldwide location. This issue has also been studied in the literature. Duffie and Beckman [23] have suggested tilt angles equal to the latitude for maximizing the annual yield. Tang and Wu [24] and Handoyo, Ichsania, and Prabowo [25] have recently proposed another criterion.

The energy losses by the tank (to the ambient for aboveground tanks, or to ground for underground ones) along the *n-teenth* month, $E_{amb}{}^n$, is calculated in Eq. (10) by integrating the Q_{amb} power term during its time step (Δt_n), and by considering the tank temperature of the previous month, T_{n-1}. In Eq. (10), λ and s are, respectively, the thermal conductivity and thickness of the insulation material, and A is the overall external area of the tank.

$$E_{amb}{}^{n} = (\lambda/s)A\left(T^{n-1}–T_{a}{}^{n}\right)\Delta t^{n} \tag{10}$$

Remembering that the monthly averages of energy consumed by the space heat system (E_{heat}) and the monthly mean ambient temperatures are given by **Table 1**, now all terms of Eqs. (4)–(6) can be explicated. So, these equations can be solved by the iterative procedure described before, by:

$$M\,c\,\left(T_2{}^1 - T_1{}^0\right) = E_{solar}{}^1–E_{amb}{}^1–E_{heat}{}^1 \tag{11}$$

$$M\,c\,\left(T_3{}^1 - T_2{}^1\right) = E_{solar}{}^2–E_{amb}{}^2–E_{heat}{}^2 \tag{12}$$

$$M\,c\,\left(T_1{}^1 - T_{12}{}^1\right) = E_{solar}{}^{12}–E_{amb}{}^{12}–E_{heat}{}^{12} \tag{13}$$

Let us discuss now the accuracy of this numerical methodology. There are two numerical approximations introduced by this explicit one-step scheme, related to the using of the temperature at the previous month (T_{n-1}), instead of using (T_n) for calculating the heat losses to the ambient (Eq. (10)) and the collector's efficiency (Eq. (9)). These two approximations together with the monthly approximation of the temperature rate (dT/dt), can be improved by reducing the time step, which is similar to any other numerical solver. This way, for all cases studied here, is provided three numerical codes (see *Complementary Material* in our Mendeley Dataset, [20]. The first code (namely 12 months) is based on a monthly time step (Eqs. (11)–(13)). The second code (namely 365 days) follows the same physical approximations that the previous one, but based on a daily time step and so, it gives us a more accurate solution. As it will be discussed in the next section, the major improvement achieved is related to the calculation of efficiency (Eq. (9)), which on the monthly modeling has tended to underestimate the efficiency during the winter months, which in turn penalizes the system's performance. In addition, there is a fourth approximation "hidden" in our numerical modeling, which is related to the calculation of the α_n factors. This approximation is programmed on both previous codes by using the same monthly values calculated, as

	Monthly heating demand		Ambient mean temperature
	%	kWh	(°C)
January	28.6	4,521	−12
February	11.9	1,884	−8
March	3.4	533	−2
April	0.9	148	5
May	1.8	281	10
June	—	0	13
July	—	0	17
August	—	0	18
September	—	0	12
October	—	0	5
November	16.3	2,569	3
December	37.1	5,855	−9
Annual	100%	15,795	4.5

Table 1.
Monthly fractions of heating demand and ambient mean temperatures for Okotoks.

was previously described by using our solar (solar_trajectory) code [20]. This code calculates the α_n factors going through the daily sun apparent trajectory by using a time step of 0.1 hours. Therefore, here is also provided with a third code (namely full365) that includes a daily calculation of these α_n factors. These calculations can be performed day by day going through a very time-consuming task, by using our solar code (solar_trajectory). Fortunately for the reader, this procedure has been already programmed by using a Visual Basic subroutine that is provided too (clicking the right button of your mouse over the name of the sheet and then, selecting the option 'view code'). Here, is also provided with a second sheet (namely 0.01 h) that includes a more accurate (taking 0.01 h time step) calculation for solar_trajectory.xls, in order to minimize the error of this procedure too. Hence, now the accuracy of our first monthly model (12 months) can be estimated by comparing its performance against this most refined daily model (full365) developed. According to the observation that this last model provides always solutions with very small variations of the main variable (that is, the tank temperature varies down 0.3°C in every daily step), the numerical error of this model can be estimated as negligible and thus, the total error of our first monthly model can be estimated by comparing against this last code; this way, we have observed always solutions within the +/− 10% bandwidth error.

3. Results

3.1 Results for our previous design

Let us summarize now the major results obtained by using our previous design, which is based on the present design developed here. So, that design is similar to the present design, but has some minor differences:

1. The water tank is heated up to 85°C (instead of 120°C);

2. The system is designed to only satisfy the space heating demand (instead of including other demands);

3. It is not used standard heaters as a backup system.

This design was performed in our previous work [1]. Summarizing, the analysis performed had found several different behaviors:

a. For small tanks, the system performance is better as much as the tank size is increased. This expected behavior has led to traditional projects using very large STES systems.

b. For tanks larger than a certain size, the opposite behavior is found, that is, the system performance is worse as much as the tank size is increased. In this case, was observed that the drawback of larger heat losses (due to the larger tank area) counterbalance the positive effect of having a larger heat storage capacity.

c. For small tanks, it is mandatory to maximize the collector yield during winter, and so, very high collector's tilt angles must be used, like 78°. It also was observed that this high tilt could be obtained by mixing some collectors on a more common tilt angle (like 45°), and others put onto vertical walls (90°).

d. Otherwise, for large tanks (that works properly as a seasonal storage system) is not mandatory to use high tilt angles, and lower angles (usually used for maximizing the annual yield, like 45°) can be used.

e. The results obtained for vacuum-tube solar collectors are noticeable better than for flat ones. Flat collectors get poor average efficiencies and are almost null during winter. Meanwhile, vacuum-tube collectors can obtain some interesting yield during winter, even on cloudy days.

f. These thermal performances are related to their economic performances. It is possible to fulfill the heating demand by using a small (72 m^3) aboveground water tank with 18 vacuum-tube collectors (solar area 37 m^2), costing about 30,500 euros. On the other hand, for flat solar collectors it is required to install a larger tank (170 m^3) and 23 flat collectors (solar area 48 m^2), and so, the overall cost increases to 45,400 euros if an aboveground tank is installed. However, maybe this large tank causes an undesirable visual impact and would be preferred an underground siting, in which case the overall cost increases to 112,500 euros.

The calculation of any solution implies choosing a tank size (M) and calculating the minimal number of solar collectors (N) needed in order to satisfy the space heating demand, that is, that the water temperature of the tank works always within the usable range (33–85°C) in order to provide the space heating demand. However, in this procedure, we must keep in mind that the dynamical model considers only average parameters (temperatures, etc.). So, during very cold days and especially when very small tanks are chosen, this calculation could not be conservative, since the thermal storage capacity of the water tank could not be enough for overpassing such an event. Therefore, let us study now the behavior of the STES system during the worst weather event (having several fully cloudy days) that we will define as a ten-day cloudy-weather event. This event is calculated in our model by means of not considering the average monthly solar irradiance, and instead considering the collector's yield obtained during cloudy days. So, we are calculating the (higher) number of collectors needed for solving this extreme condition. These N numbers are illustrated in **Table 2** (in brackets) for φ = 78°, together with the usual average solution. Here is observed that this very cold winter leads to very poor performances for small tanks, which must be supported by using much more collectors. Otherwise, large tanks can easily manage this scenario; for example, case **A** (M = 170 m^3) provides enough storage for one month, and so, the number of collectors are the same in both (average and worst) cases.

Table 3 summarizes the breakdown of cost (for the case φ = 78° and ten-day storage capacity) for the previous cases studied in our previous work [20]. It is interesting to note that the larger tank (case **A**) obtains a total cost slightly higher than the other ones, but it provides the largest storing capacity (one month) that can fully provide the heating demand. However, in this case, it should also be evaluated the visual impact of placing this large aboveground tank close to the house. So, let us study now another option for providing this large storage capacity, which is performed by using the next smaller tank (case **B**) with eighteen solar collectors, so getting a total cost of €32,200.

Table 4 repeats the previous analysis by using flat solar collectors (STES_Okotoks-Flat collectors, in [20]) having each one the same solar area (2.088 m^2) as the previous vacuum-tube collectors, but, of course, they both have different efficiency curves,

M (m^3)	N 78°(#)	$\eta_{collector}$ (78°)	N 45°(#)
1,360	8(8)	69%	8
170-**A**	8(8)	63%	9
72-**B**	16(18)	59%	19
21-**C**	27(34)	58%	33
4,6-**D**	30(76)	57%	36

Table 2.
(N, M) solutions for vacuum-tube collectors.

	A	B	C	D
Collectors	€4,800	€10,800	€20,400	€45,600
Tank	€20,031	€11,362	€5,038	€1,849
EPS insulation	€6,290	€3,515	€1,554	€555
Siting preparation	€68,000	€28,800	€8,400	€1,840
Fixed costs	€4,344	€4,344	€4,344	€4,344
Total (underground)	**€103,500**	**€58,800**	**€39,700**	**€54,200**
Total (aboveground)	**€36,400**	**€30,500**	**€31,600**	**€52,400**

Table 3.
Breakdown of costs for different solutions (φ = 78°).

M (m^3)	N 78° (#)	$\eta_{collector}$ (78°)	N 45°(#)
1,360	13(13)	54%	14
573	12(12)	48%	12
170-**A**	23(23)	28%	28(28)
72-**B**	51(76)	23%	61(90)
21-**C**	62(211)	22%	76(239)
4,6-**D**	62(∝)	22%	75(∝)

Table 4.
(M, N) solutions for flat collectors (tilt angles 78° or 45°).

according to **Figure 1**. The performance of these flat collectors has been simulated (following our previous discussion) by reducing 25% the solar factors α^nxG^n previously calculated for vacuum-tube collectors. Similarly, the I_n fluxes previously calculated are now 25% reduced and then used in the efficiency equation (Eq. (9)).

Now, by comparing **Tables 2** and **4**, we can observe that flat collectors always get lower efficiencies than vacuum-tubes ones and that their efficiency decreases strongly as much as the tank size is reduced, and so, their working temperature is increased. Another difference regarding vacuum-tube collectors is that flat collectors achieve negligible efficiencies during winter and so, their performance is highly penalized when small tanks are used. Otherwise, it is interesting to note that their performances are very reasonable by using large tanks, in which they can take advantage of their higher efficiencies during summer (**Table 4**).

In order to estimate the total cost of these alternative designs for Okotoks, it will be assumed that the unitary cost of flat collectors is equal to previous vacuum-tube ones, regarding that there is a wide range of commercial models for both kinds of collectors and so, different cost choices. **Table 5** shows the total investments for the previously studied cases. Here is can be observed a different behavior regarding the previous systems with vacuum-tube collectors, since now the best choice is always obtained with the largest tank (case **A**). In addition, this large tank can support the collectors installed on different tilt angles.

It is interesting to compare this case (solar area 37 m^2) with the Okotoks' project that uses a similar collector's area (44 m^2 per house). This 170 m^3 tank gets an average efficiency of 82%, which is higher than the Okotoks STES efficiency (60%), due to their lower size and higher thermal insulation (the Okotoks reservoir uses 583 m^3 of rocky underground per house). The comparison with the Friedrichshafen's project is also interesting since both use water tanks as STES system. Here is observed that the German project uses a larger tank (320 m^3) with low efficiency (60%) and a higher

Tank siting -tilt angle	A	B	C
Underground-78°	€112,500	€93,600	€146,000
Aboveground-78°	€45,400	€65,400	€137,800
Underground-45°	€116,400	€102,600	€163,000
Aboveground-45°	€48,400	€73,800	€154,600

Table 5.
Total costs for flat collectors.

solar area (108 m^2) too, leading to a noticeable higher overall cost (128,000 U$) per equivalent Okotoks' house.

Finally, **Table 5** is presented the total cost of these cases for underground tanks, calculated from the Galway's underground tank (€17,600). Here we can observe that an underground tank always leads to a remarkable higher cost than the aboveground option. Hence, since a small tank can be conveniently be insulated, we will prefer aboveground tanks. Moreover, as we will discuss in the next section, this aboveground could be installed within the greenhouse near the house, and this way, its heat losses can be useful for warming the greenhouse.

3.2 Results for the new design

The new design takes advantage of four main concepts:

1. The utilization of other secondary demands of heat throughout the whole year in order to fully exploit the collector yield. Regarding that space heating demand is very concentrated during four months in cold locations as Okotoks (from November to February, 94%), it was observed in the previous study [1] that the small tank cannot store the solar production along the year, and so it must be avoided during seven months (being needed just one month for heating the tank in advance to cold season). Hence, here is exploited this surplus of heat to fulfill other demands, like warming a greenhouse (during spring and autumn) and swimming pool (during summer).

2. Here is allowed to help the solar production during winter (the most exigent demand) by using small standard electric heaters (consuming total energy down 10% of annual solar production). This backup system helps us to downsize the STES (the main cost), as we shall see in this section. This choice is a smart economical optimization since these electric heaters only must work during the extremely cheap (about 0.1 euros per kWh) valley tariff.

3. The capability of a commercial stainless steel water tank for withstands slight overpressures (up to 3 bars). Hence, in this design is overheated the water tank up to 120°C (2 bars) and this way, the usable heat (from 120–33°C) is doubled regarding previous design.

4. The low cost of these tanks that are massively manufactured for daily-storage solar & heat pump applications (space heating and sanitary hot water). These tanks are available up to 5,000 liters costing about 3,000 euros, and including high-quality thermal insulation, internal double heat exchanger, and thermostat and standard electric heater. So, we can build the whole STES system (by adding a water pump and controlling unit, adding 2,000 dollars) easily by installing one or more tanks.

This novel design was developed on new software [26] that is similar to our previous dataset but includes the calculation of the secondary demand and other minor changes. The managing strategy followed here combines different objectives along the year:

1. It maximizes the secondary production from spring to autumn, by setting the tank temperature at 30°C during this warm period. So, the secondary demand is

calculated every day as the one required in order to get an equilibrium balance of energy (that is, the fully heat production from solar collectors is used as secondary demand). This low temperature (30°C) was set according to fulfill the main secondary demands considered (warming a greenhouse and swimming pool). Actually, this 30°C level is not enough for providing sanitary hot water demand (another secondary demand considered), but this last demand (calculated as 200 liters of 40°C-heated water, or 9.3 kWh per day) is almost neglected (about 1%) compared to the total secondary demand produced every day. So, the production of sanitary hot water would not change the energetic balance calculated here; maybe it implies some minor complexity (another 200 liters tank heated up to 45°C every day by solar collectors) that will not be considered here.

2. The water tank is heated up to 120°C before the peak winter demand starts (during December); it is observed that a month is enough for this purpose. Therefore during November the tank is heated by solar collectors meanwhile the space heating and sanitary hot water demands are fulfilled, and all the other secondary demands are avoided.

3. The water tank is continuously cooled during December since the demand is higher than the solar yield, but the backup electric heaters are used in order to keep the minimum usable level (33°C) on the last day of December.

4. The tank temperature is increased along January without using the backup system since the primary demand (avoiding other secondary demands) is lower than the solar yield. This way, it is observed that the tank is heated at about 70°C on the last day of January.

5. During February and March, it is considered that the "danger" condition has been overpassed (there is still some heating demand, but it is remarkable lower than the previous one). So, our strategy for this period consists in setting the water tank to 45°C, in order to provide some margin for any "bad weather event" that might occur. Therefore, the system could provide another secondary demand (to warm the greenhouse), starting with a "heat punch" that is calculated in order to get the 45°C desired level.

6. From April to October (the warm season), the tank is set to 30°C (starting again with a "heat punch"), and all solar production is available for other secondary demands.

Following this managing strategy, **Table 6** shows the results obtained by a sensitivity analysis on the number of collectors (N) for 10 m^3 water tanks (that could be built by using two 5,000 liters commercial tanks). All cthese cases are described in our Dataset [26], cases #1 to #5. In all these cases is calculated the annual heat production from solar collectors, E_{solar}, and the annual heat production from electric heaters, E_{elec} (it absolute value and percentage of E_{solar}), and it is also noted the continuous electric power, P_{elec}, that would be required during the valley tariff (8 hours per day), and the number of months of the year that the system could provide these secondary demand (warming a greenhouse or swimming pool). Let us note here that the total fixed demand to fulfill (space heating, 15,795 kWh/y and SHWD, 3,407 kWh/y, totalizing

N (#)	E_{solar} (kWh)	E_{elec} (kWh)	E_{elec} (%)	P_{elec} (kW)	Months 2nd heat
10	53,760	6,424	11.9	15	8.0
15	74,259	3,664	4,9	11	8.6
20	94,363	1,852	2.0	7	9.0
25	117,243	768	0.7	3	10.6
30	139,889	99	0.1	0.4	11.0

Table 6.
Sensitive analysis for number of collectors, N, *and backup heaters* (M = $10\ m^3$@120°C).

19,202 kWh/y) is several (among 3 and 7) times smaller than the total energy produced by including these secondary demands too. In all these cases, the thermal efficiency of the STES system is very high (around 95%), as much as the average collector efficiencies (around 67%). Although the tank must be overheated during winter, and in this condition, the collector efficiencies are down to 40%, these annual efficiencies are high because most parts of year the collectors (and tank) work on low temperatures.

The sensitive analysis performed in **Table 6** is now related to cost analysis (**Table 7**), by considering each 5,000 liters tank (€4,000), the auxiliary systems (controlling system and pump, €2,000), and each 20-tubes (2.088 m^2 solar area) collector (€500, similar to our previous work). The cost of electricity is always considered as 0.1 €/kWh, according to the valley tariff, for the backup system and for calculating the annual saving obtained compared to standard system (fully providing heating by electrical heaters).

Table 7 shows that higher the number of collectors is lower the payback period is, although with slight differences (10%) above twenty collectors. So, the optimal solution could be 20 to 30 collectors, according to the investment desirable and the secondary heating demands that actually are required. This last point is remarkable; the previous analysis is based on considering that the fully solar production is utilized, otherwise, the cost optimization would noticeably change. For example, let us consider now the opposite behavior, that is, without others' secondary demands (except SHWD). In this condition, the total heating demand is 19,202 kWh/y and so, the maximum annual saving achievable is €1,920 (minus the backup consumption). So, by calculating again the payback period for this condition (the last column in **Table 7**), are obtained values from 11.3 to 13.1 years, being the optimal around 15 collectors. Hence, we can conclude that an optimal point for every condition is around 20 collectors.

N (#)	Total cost (€)	Backup (€/y)	Saving (€/y)	Payb. (years)	Payb.* (years)
10	15,000	642	4,700	3.2	11.7
15	17,500	366	7,100	2.5	11.3
20	20,000	185	9,300	2.2	11.5
25	22,500	77	11,600	1.9	12.2
30	25,000	10	14,000	1.8	13.1

** represents the payback period without secondary demands*

Table 7.
*Cost analysis for previous (**Table 6**) cases.*

M (m^3)	η_{tank} (%)	E_{solar} (kWh)	E_{elec} (kWh)	E_{elec} (%)	P_{elec} (kW)
10	95	94,363	1,852	2.0	7
50	89	94,704	744	0.8	3
100	82	91,354	0	0	0
100*	71	78,721	0	0	0

** represents the payback period without secondary demands*

Table 8.
Sensitive analysis for tank size (N = 20).

Let us study now the sensitivity analysis about tank size, M. It is possible that a larger tank could obtain a better performance since the backup system would be less required. This is true, but it must be counterbalanced with higher heat losses (due to the larger tank area), and higher costs as well. **Table 8** shows this effect (for $N = 20$) by considering $M = 10$, 50 and 100 m^3. This last case is repeated (*) for considering a slightly different strategy; in this, the large storage capability is exploited for collecting energy during the summer (this tank can store about 42 days of winter heating demand), what could be useful is the dweller does not have a swimming pool, in order to use this stored energy during the spring and autumn seasons. This way, the usable season for the greenhouse could be started before (January) and ended after (October) that is, extending it two months. **Table 8** shows that a larger tank suffers larger heat losses that overpass the benefits of having a larger storage capacity (that is, the overall energy production achieved in this way is lower). Besides, the total cost related to a larger tank is increased noticeably, being €52,000 and €92,000 for M = 50 and 100 m^3, respectively.

Finally, all the previous analyses show that the solar, thermal and economical behaviors are strongly linked. Hence, simple explicit modeling as it is performed here has been demonstrated to be useful for optimizing altogether the system parameters.

4. Conclusions

In this work was studied the performance of solar + STES systems based on many vacuum-tube solar collectors and a small well-insulated aboveground water tank, which is used to provide all the heat demands related to a single-family house in cold climates. This approach is innovative in many manners. These kinds of systems have been traditionally designed to fulfill the space heating demand of many houses together in cold climates that are concentrated during winter, but in this case, it is also designed to satisfy other secondary demands of dwellers along the year, like sanitary hot water, and to warm a greenhouse (from spring to autumn) and a swimming pool (during summer). Besides, the traditional approach followed in most projects has used many flat solar collectors with a huge STES that provides seasonal storage. On the contrary, here is proposed to use many vacuum-tube collectors and a short-term STES, which provides a solution with noticeably lower costs.

This work has discussed the radical differences between both designs from a designer point of view, that is, to perform "inverse engineering" (from results to design), in order to understand the motivations behind each design. It has shown that there are many hidden concepts supporting the traditional design. So, the choice of a huge STES seems to be motivated by the expectative about reaching lower costs and heat losses, due to scaling up the reservoir size. As it was discussed here, none of both issues has been actually achieved in present large projects. Firstly, it is true that the volume/area ratio can be reduced by enlarging the tank size, which could lead to getting lower heat losses and costs as well. However, this effect is actually overcome by higher heat losses caused by the fact that is not possible to put thermal insulation under a huge and overweight tank. For example, the Friedrichshafen's project uses a 12 m-height underground tank (walls built by 30 cm-thickness reinforced concrete and a stainless steel 2 mm liner) in which there is no insulation on its bottom third part, and it achieves overall heat loses about 40%, similarly to the Okotoks' project on its huge heat reservoir built by deeply drilling the rocky ground. Secondly, the cost of building a huge (12.000m^3) tank as the Friedrichshafen's project uses, is noticeably increased by the requirements of mounting it within an underground site, since such as huge tank would cause a high visual impact if it is mounted aboveground. Furthermore, we have already discussed in the previous work that the ultimate motivation behind the use of a huge heat reservoir is to support the utilization of flat solar collectors. This kind of collector cannot give yield during winter (when the space heating demands occur) in cold climates; so, this choice obeys us to consider a seasonal STES, in which the flat collectors accumulate heat during summer.

On the other hand, the novel design proposed here uses many vacuum-tube collectors, which can obtain a remarkable yield during winter. This way, this solar system can be supported by a short-term (providing down one month of the heating demand) STES system, which in turn reduces noticeably the overall cost. This way, this short-term STES can be performed by using an aboveground stainless steel water tank, which can be easily wrapped with thermal insulation in order to achieve overall heat losses of about 5%, and achieving overall cost remarkable lowers that the traditional design.

According to the performance of both designs, the traditional design and novel one proposed here, we can point out that the preference for flat collectors is the primary cause behind the unaffordable costs achieved by all projects developed up today. We guess that this issue has been overlooked in previous analyzes, but we want to be clear about this. There are many customers reluctant to put vacuum-tube solar collectors in their homes. This is true especially in Europe, where is forbidden to install collectors that waste water from the distribution grid. This situation can occur (mostly in summer vacancies, that is, without hot-water consumption) for vacuum-tube collectors. In this case, these collectors can suffer a dangerous overheating solved by discharging steam to the ambient. This solution could be acceptable for use as a second (security) system, but this is completely unacceptable for using periodically (that is, working actually as a controlling system). For example, in the event that it happened that this pressure-relief valve gets stuck and the overpressure cannot be released, the water tank could suffer a catastrophic rupture, which nobody wants to occur in his home. Perhaps, this weakness of the design of vacuum-tube collectors is actually the major limitation for their massive application. It is funny, but this overheating is cause for their successful improvement in getting lower heat losses (achieved by using better sensitivity coatings with lower infrared emissivity), as was shown in a recent work. In this work is discussed how this drawback could be solved by just making a step back in the development of better sensitive coatings [18]. This solution is affordable and can be easily applied by manufacturers, instead of the complex and expensive solutions that are currently under development, which propose smart selective coating with temperature-controlled solar light transmittance [27, 28]. Moreover, in this work, Juanicó also proposes to enlarge the number of vacuum tubes and the size of the water tank of the average collector (about 40 tubes and 500 liters water, instead of the average 20-tubes 200-liters collector) in order to noticeably enhance the capability of the solar collector for providing the hot water demand during several cloudy winter days, as well as this design noticeably reduces the risk of overheating. This new design of collector intends to overcome the present limitation of solar collectors that, at the present, satisfy only partially the average dweller demand.

According to this last design, we can realize now that the small (solar + STES) system proposed here follows this concept. A relatively large tank size (10 m^3) can be enough large to overcome concerns about overheating during vacancies. Moreover, the thermal–hydraulic configuration used here (in which the heat produced by solar collectors is transferred to the tank only when the controlling system does that) forbids the risks of overheating at the tank. Besides, the high-temperature (up to 120° C@2 bar) heat reservoir proposed here helps to overcome this concern, because the thermal efficiency of commercial vacuum-tube collectors decreases noticeably working at this temperature. These features altogether should convince us to use vacuum-tube collectors as a feasible and safe option.

This work has studied the advantage of using a STES that can withstand higher temperatures (up to 120°C). This level is higher that the temperature used in previous projects (up to 85°C), but this novel proposal could be easily performed by using one of more commercial stainless steel tanks (5,000 liters) that are manufactured at low cost and including all the auxiliary systems needed: two heat exchangers built by copper coils, standard electrical heater, pressure relief valve (3 bar), and good-quality thermal insulation. So, this design exploits the advantage of using low-cost commercial tanks manufactured by Chinese factories, mostly for their solar internal market. We can conclude that this novel proposal could be a "silver bullet" useful in order for this technology can become an affordable and suitable solution.

Up today, this solar+ STES technology remains within the under-developing prototypical level after more than twenty years of studying and a similar number of large-scale projects tested (mostly in German). Moreover, which is worst, I think, is the fact that there are negligible chances of reaching success in the future, since the cost of a huge STES system could hardly become enough cheaper to become a technology economically competitive. Moreover, I think we cannot expect a good prospective for this technology in next years, since also the price of solar collectors seems to have reached a steady level after reaching a large massive production scale. On the other hand, during this period the photovoltaic panels have noticeably become cheaper, as well as other technologies related to the production of electricity and its utilization for heating water, such as 1) the generalization of net metering and distributed generation from homes; 2) the reduced price of battery backup systems, by the hand of the generalization of electric cars that drives the growing up of the second-life battery market; 3) the generalization of air-water heat pumps, which are useful for providing all these low-temperature demands of heat having superlative efficiencies (up to 400%), or conversely, this is equivalent to increase four times the electricity from PV panels.

The vacuum-tube collectors can obtain significant yields during winter, even during cloudy days [29]. Therefore, by using many vacuum-tube collectors the winter demand can be fulfilled working with a short-term STES system. This design is noticeably cheaper than the traditional one based on a huge tank, according to the lower cost of a small tank. Besides, this work will be also studied the thermal and cost performance achieved when this small tank is installed aboveground, instead of the traditional underground siting used in large projects. Hence, it was demonstrated that by using reasonable thermal insulation, the heat losses of the aboveground tank are similar that the underground one, but, since the aboveground tank has an overall cost noticeably lower (up 4 times) than the underground one, the aboveground choice is preferred here.

Finally, this work was a study of the economical optimization of these systems by adding a partial generation of heat from standard electrical heaters. This configuration is reasonable because it could take advantage of using the very low cost "valley" tariff during the night (11 pm to 7 am) for household dwellers.

Author details

Luis E. Juanicó
Instituto Andino Patagónico en Tecnologías Biológicas y Geoambientales (IPATEC), National University of Comahue (UNCO) and National Council of Scientific Researches (CONICET), Argentina

*Address all correspondence to: juanico@comahue-conicet.gob.ar

References

[1] Juanicó LE. Heating houses by using vacuum-tube solar collectors and a small above-ground water tank: A cost-effective solution for maritime climates. Advanced in Building Energy Research. 2020. DOI: 10.1080/17512549.2019.1688186

[2] Novo AV, Bayon JR, Castro-Fresno D, Rodriguez-Hernandez J. Review of seasonal heat storage in large basins: Water tanks and gravel–water pits. Applied Energy. 2010;**87**:390-397

[3] Rad FM, Fung A. Solar community heating and cooling system with borehole thermal energy storage: Review of systems. Renewable and Sustainable Energy Reviews. 2016;**60**:1550-1561

[4] Sibbitt B, McClenahan D, Djebbar R, Thornton J, Wong B, Carriere J, et al. The per- formance of a high solar fraction seasonal storage District heating system: Five Years of Operation. Energy Procedia. 2012;**30**:856-865

[5] Eto JH. On using degree-days to account for the effects of weather on annual energy use in office buildings. Energy and Buildings. 1988;**12**:113-127

[6] Saic (Science Applications International Corporation). Drake Landing Solar Community Energy Report 2008–2009 (CM002171). 2010. Retrieved from https://www.dlsc.ca/reports/bjul15/DLSC_2008- 2009_Annual_Report_v3.0.pdf

[7] NRC (Natural Resources Canada). Backgrounder. Drake Landing Solar Community Hits the Target for Year Three Performance. 2010. Retrieved from: https://www.nrcan.gc.ca/media-room/1505 and https://www.dlsc.ca/

[8] Bauer D, Marx R, Nußbicker-Lux J, Ochs F, Heidemann W, Müller-Steinhagen H. German central solar heating plants with seasonal heat storage. Solar Energy. 2010;**84**(4):612-623

[9] Schmidt T, Mangold D, Müller-Steinhagen H. Central solar heating plants with seasonal storage in Germany. Solar Energy. 2004;**76**(1–3):165-174

[10] Colclough S, Griffiths P. Financial analysis of an installed small-scale seasonal thermal energy store. Renewable Energy. 2016;**86**:422-428

[11] Colclough S, Griffiths P, Smyth M. Solar energy storage- critical success factors for passive houses in Ireland. In: World Renewable Energy Congress XI. Proceeding edited by The World Renewable Energy Congress, U.K; 25–30 September 2010, Abu Dhabi, UAE. 2010. Available at: https://pure.ulster.ac.uk/en/publications/solar-energy-storage-critical-success-factors-for-passive-houses-

[12] Ucar A, Inalli M. A thermo-economical optimization of a domestic solar heating plant with seasonal storage. Applied Thermal Engineering. 2007;**27**:450-456

[13] Ucar A, Inalli M. Thermal and economic comparisons of solar heating systems with seasonal storage used in building heating. Renewable Energy. 2008;**33**:2532-2539

[14] Incropera F, Dewitt D. Fundamentals of Heat and Mass Transfer. 7th ed. New York: John Wiley; 2011

[15] Bowa. Heat Exchangers. 2017. Retrieved from: https://bowasolution.es.aliexpress.com/store/1333457

[16] Skiba E. Understanding Solar Collector Efficiency. Contribution Letter in the Apricus, Vacuum- Tube Solar Collector Manufacturer's Website. 2018. Retrieved from: http://www.apricus.com/upload/ userfiles/downloads/Apricus-article-understanding-solar-collector-efficiency.pdf

[17] ESTIF, European Solar Thermal Industry Federation. 2006. Retrieved from: http://solarprofessional.com/sites/default/files/articles/ajax/docs/6_SP2_6_pg58_Stickney-8.jpg

[18] Juanicó LE. Modified vacuum tubes for overheating limitation of solar collectors: A dynamical modeling approach. Solar Energy. 2018;**171C**: 804-810

[19] Manufacturer Website. 2018. Retrieved from: http://www.soltigua.com/itracker/

[20] Juanicó LE. Modeling of seasonal thermal energy storage systems (STESs) with solar collectors. Mendeley Data. 2019;**V2**. DOI: 10.17632/c535555vfy.2

[21] Juanicó LE, González AD. Calefacción solar en edificaciones con acumulación en gran reservorio de agua. Revista INVI. 2018;**33**(93): 153-172

[22] NASA. NASA Website. 2018. Retrieved from https://power.larc.nasa.gov/. NASA Prediction of Worldwide Energy Resource

[23] Duffie JA, Beckman WA. Solar Engineering of Thermal Processes. 2nd ed. New Jersey, U.S.: John Wiley & Sons, Inc.; 1991

[24] Tang R, Wu T. Optimal tilt-angles for solar collectors used in China. Applied Energy. 2004;**79**:239-248

[25] Handoyo EA, Ichsania D, Prabowo. The optimal tilt angle of a solar collector. Energy Procedia. 2013;**32**: 166-175

[26] Juanicó LE. Modeling of Seasonal Thermal Energy Storage Systems (STESs) with solar collectors NEW VERSION. Mendeley DataBase. 2021 under production

[27] Mercs D, Didelot A, Capon F, Pierson JF, Hafner B, Pazidis A, et al. Innovative smart selective coating to avoid overheating in highly efficient thermal solar collectors. Energy Procedia. 2016;**91**:84-93

[28] Muehling O, Seeboth A, Ruhmann R, Eberhardt V, Byker H, Anderson CD, et al. Solar collector cover with temperature-controlled solar light transmittance. Energy Procedia. 2014; **48**:163-171

[29] Sabiha MA, Saidur R, Mekhilef S, Mahian O. Progress and latest developments of evac- uated tube solar collectors. Renewable and Sustainable Energy Reviews. 2018;**51**:1038-1054

Chapter 11

Highlighting the Design and Performance Gaps: Case Studies of University Buildings

Karishma Kashyap, Usha Iyer-Raniga and Mary Myla Andamon

Abstract

Buildings are one of the highest emitters of greenhouse gases globally. To reduce the detrimental effects of buildings on the environment and recognise their potential for emissions reductions, a transition towards sustainable building solutions has been observed globally. This trend and the associated benefits have been discussed and argued for more than three decades now. However, the impacts of sustainable buildings are yet to be demonstrated at macro, meso, and micro levels in the community, as the actual versus expected performance of such buildings are still being questioned. Consequently, this entry discusses the concepts underpinning sustainable buildings outlining the drivers and practices to achieve sustainable built environment solutions from the design to operation stage using university buildings as a case study. The chapter also recommends evidence-based solutions on understanding the actual and perceived gaps to achieve expected performance using "Green Star" rated academic buildings in Australia.

Keywords: facility management, green academic buildings, green construction, green star, post-occupancy evaluation, sustainable buildings

1. Introduction

Globally, it is a critical time to act on climate change either through developing appropriate policies, focused research or implementing sustainable and green strategies across various sectors. The report by United Nations Environment Programme (UNEP) and International Energy Agency (IEA) [1] indicates that by 2056, globally, there is going to be a rapid increase by five-fold in the economic activity and by three-fold in energy consumption compared to the current numbers. Furthermore, there would be a population increase by 50% and a drastic rise in global manufacturing activities imposing further pressure on the resources and planet. Buildings alone are responsible for 40% of the GHG emissions [2, 3]. As a consequence of the rapid growth in economic activities, urbanization, and growing population, the impact of the built environment sector on the planet is anticipated to worsen further. The building industry's exhaustively growing consumption of energy through the construction to operation phase is considered as a major contributor to environmental pollution, producing an enormous amount of waste [4–8].

However, the building industry also demonstrates high potential and has capabilities to contribute towards significant emissions reduction [9–10]. As stated by the United Nations Environment Programme [2], "the solution that sits at the intersection of urbanisation and climate change which can cut carbon emissions, boost productivity, and enhance the health and wellbeing of people is a green building." The outcomes and commitments of the built environment sector have also been recognized as a critical sector at COP26 at Glasgow demonstrating leadership, resilience, and social inclusiveness for climate action [11]. However, a "sustainable" or "green building" is a broad and complex concept, which is still quite perplexing for the building industry. Sustainability is a concept that is built upon three key aspects, namely environmental, economic, and social sustainability. These three aspects need to be incorporated collectively and cannot be targeted individually to achieve expected results [4, 6, 7]. Jenson [12] states that it is when all these aspects fit together in the planning and governance strategies, one can deliver buildings that perform as desired and are truly sensitive to our environment. This is a challenge continually encountered by design, construction, and building management practitioners. The objectives primarily established are mostly only incremental changes; however, what is required is a step-change in these design and management approaches being currently followed to acknowledge the veritable potential of buildings.

Supporting this concept, many recent studies have argued the potential of buildings for delivering significant cuts in emissions at no cost or little extra cost, reducing harmful environmental effects, and acting as the backbone of any economy [3, 6, 8]. There is immense potential for this as the building industry employs more than 10% of the global workforce [1, 3, 6, 10], resulting in gains in social outcomes as well. Underpinned by this argument, this chapter discusses the key concepts of sustainability in buildings and recommends a best practice model to close the gap between actual versus expected performance taking "Green Star" rated tertiary education institutions in Melbourne, Australia as case studies. Green Star is the internationally recognised sustainability rating and certification of the Green Building Council of Australia (GBCA). This is cross-referenced in Section 2 – Background (p4) with further information. The results of the study presented assists the design community to deploy strategies that not only meet a building's energy and savings targets but also respond to occupants' requirements of space comfort and use, creating an overall efficient system for users and building managers alike, thus achieving all three components of sustainability – reduction of environmental impacts, increased social outcomes, and reduced economic costs.

2. Background

The two aspects of the overall building purpose and operation are not necessarily always in sync as has been demonstrated to date by the way the industry has approached this problem in the form of split incentives [13]. Many organisations feel that just complying with the environmental norms or national regulations is enough to meet user needs, and their work is completed once they manage to deliver a building as per such guidelines. Examining the performance of a building is also mostly restricted to evaluating the operation of a building with respect to energy consumption and energy efficiency. Most initiatives taken by facility managers and senior management are to ensure that the utilities, that is, electricity, gas, and water consumption of the building meet the standards overlooking other key concepts of performance management, such as incorporating the occupant's perception and outlook in building management, performance evaluation through a building's life cycle, and other such social considerations beyond the environmental. However, a building in operation is a complex system and its management should not be restricted only to the knowledge of facility managers. It should focus on more integrated and holistic approaches, where the organisational objectives, knowledge of facility managers, review by the external stakeholders, and occupant perceptions should all be considered together to be satisfactory and present optimum results for these stakeholders.

As indicated earlier with green buildings, the assessments have generally been restricted to the measurement of energy being the key focus for most organizations when aiming to reduce the emissions or overall carbon footprints. When including other aspects of building performance, many studies [14–19] have discussed the significance of monitoring design and other internal building features, with a focus primarily on indoor environmental quality (IEQ) and thermal comfort conditions. However, not much focus and practical exploration have been undertaken to consider occupant satisfaction for evaluating overall building performance and closing the loop between the occupants and building management to achieve holistic green outcomes [20, 21].

Tertiary education institutions have a critical role in the current decade to innovate, excel, and set an example in meeting the needs of the changing environment [22]. The strategic vision of universities needs to be one that accounts for the rapidly changing world, and to be flexible enough to recalibrate and refresh as conditions around them change. Universities globally are investing billions of dollars into building construction that showcases their sustainability commitment [23] designed for being used as science laboratories to incubate innovative green technologies facilitating academic and research activities. Universities are constantly upgrading to sustainable standards leading through design, innovation, and adapting to new reforms and requirements by the industry around the globe. Such reforms are formulated to cultivate a population that is knowledgeable and active towards their crucial role in sustainable development and combating climate change.

Over the last several years there has also been a shift away from a prescriptive approach to sustainable design towards the evidence-based evaluation of actual performance through life cycle assessments (LCAs). While LCAs are not yet a consistent requirement of green building rating systems and codes, there is a trend towards requiring LCAs and improving the methods for conducting them [24]. Here is where green building assessment tools play a role in guiding the design professionals towards green approaches and eliminating bad performing buildings from the building industry. Building assessment tools are in a continuous state of evolution and continue to be refined to reflect new standards and goals for achieving ever-higher levels of sustainability [24, 25]. Therefore, it is essential to investigate the most current versions of these tools to gain an understanding of the requirements that must be met to achieve optimal results.

Globally, sustainable building rating systems started about 30 years ago after the development of Building Research Establishment Environmental Assessment Methodology (BREEAM) in 1990. Since then, there is an emergence of several environment rating tools for buildings, such as Leadership in Energy and Environmental Design (LEED) in the U.S.A, Eco-Quantum in the Netherlands, Promise in Finland, Eco-Pro in Germany, Haute Qualité Environnementale (HQE) in France,

Comprehensive Assessment System for Built Environment Efficiency (CASBEE) in Japan, and Athena in Canada, to name a few. The assessment methods for these rating tools vary from the perspective of scope, structure, format, and complexity. Apart from the most commonly used rating tools, such as LEED and BREEAM, other assessment tools fall into the category of qualitative and life cycle assessments, including the Sustainable Building Tool; Green Star, Hong Kong Building Environmental Assessment Method (HK BEAM) or tools adapted to specific countries, such as LEED adapted for Canada and Australian Green Star adapted for New Zealand and South Africa. In some instances, the tools have developed into a new tool, for example, the Building Assessment Tool (SBAT) influenced by BREEAM and LEED.

Although most of the existing building rating systems are voluntary, in some countries, such as Australia, it is a mandatory requirement for minimum energy requirements of new office buildings and major refurbishments as per the National Construction Code Building Code of Australia maintained by the Australian Building and Construction Board [26]. Development and uptake of such rating schemes are promoted in Australia at a national level, as the detrimental effects of buildings account for approximately a quarter of the nation's greenhouse gas emissions and two-thirds of electricity consumption [27]. Currently, two voluntary national accreditation programs exist to measure and report on the environmental performance of office tenancies and assist in guiding best practice office fit-outs. These Australian building rating systems are mandatory requirements for minimum energy requirements of new buildings and major refurbishment [28]. The two rating systems, that is, NABERS and Green Star, currently available in Australia promote sustainable building development in the commercial space. Green Star is Australia's only national, voluntary, holistic rating system for sustainable buildings and communities [29]. The Green Building Council of Australia (GBCA) developed Green Star, which is a design-based rating system that provides no guarantee of, or commitment to, a specific level of performance and confines itself to the intent of the design and to some extent the provision of the process to achieve that intent [30, 31]. The Green Star rating system is a much broader rating scheme than NABERS and assesses both environmental and liveability factors throughout a building's lifecycle [29]. Green Star has now achieved sufficient market penetration and may be considered to be part of procurement practice by industry.

This study used Green Star rated buildings. Green Star started with three rating classifications, that is, "As Design," "As Built," and "Green Star Communities." Recently a fourth component of "As Performance" [32] has been added to the rating system, with each classification having versions as and when updated.

There are several versions of each Green-Star tool that have been updated since the tool was first created. Rating buildings as per the design and built outcome to interiors, performance, and communities, Green Star supports deploying innovative practices to optimise sustainability outcomes through an array of options [33]. All the versions of the Green Star rating scheme started with a pilot version to understand the limitations and continued refinements based on changing global, national, and industry requirements. As the evaluation of the case study buildings were conducted over the period 2014–2019, the buildings used were all Green-Star buildings certified for As Design (v1) at the start of the data collection period. A total of 86 academic buildings/projects in Australia were certified As Design (v1). Of which, 35 projects were in Victoria. 37 buildings/projects were certified following As Built (v1) with 6 projects located in Victoria.

Concurrently, what is also required is to constantly upgrade university governance models for long-term sustainability. The governance strategies for most universities are focused on shaping their respective campuses, creating innovative learning spaces, and a student precinct, which supports a rich and rewarding student experience. Universities aim at developing teaching and research facilities, which by their very nature, encourage interdisciplinary collaboration and industry engagement, and explore options for campus development to accommodate increasing scale in both teaching and research. Formulating appropriate strategic plans to incorporate such outcomes is crucial and provides a framework within which to develop, implement and modify practices, associated investments, and action plans.

3. Method

For the purpose of selecting study buildings for the research, a detailed evaluation of four recently constructed Green-Star rated academic buildings in Australia was carried out to observe their performance and management structure in reality. Four tertiary academic institutions were selected in Melbourne, Australia as case study buildings, incorporating post-occupancy evaluation techniques and appropriate project management strategies. The buildings were completed within the 2010–2014 period. The key features (Building Type, Academic Faculty, Green Star Certification Score, Volume, Area, and number of Floors) of the study buildings are described in **Table 1**. The buildings are denoted as Buildings A, B, C, and D. The results provide a model framework for design stakeholders to achieve desired building performance targets for green buildings, as they are designed to perform as per specific standards but can be applicable to conventional or non-green buildings as well, drawing upon lessons learned from this study.

The study has adopted a mixed-methods approach utilizing both quantitative and qualitative techniques. For quantitative data, survey research has been deployed. Surveys have been used to evaluate occupant satisfaction levels in the case study buildings.

Currently, Building User Survey (BUS) and Building Occupants Survey System Australia (BOSSA) are the only two officially accredited post-occupancy evaluation (POE) instruments within the National Australian Built Environment Rating System (NABERS) for commercial buildings used in Australia [34]. These surveys are robust and accessible Australian alternatives to other surveys currently in use by NABERS and

Study Buildings	Building Type	Faculty	Green Star Certification (Education Design v1)	Volume	Gross Floor Area	No of Floors
Building A	New Build, Refurbished	Architecture and Design	Certified – 5 Star Green Star, 2011	16,040 m^2	13,000 m^2	10
Building B	New Build	Design and Manufacturing	Certified – 6 Star Green Star, 2013	21,000 m^2	10,000 m^2	11
Building C	New Build	Construction	Certified – 5 Star Green Star, 2014	17,500 m^2	11, 861 m^2	3
Building D	New Build	Architecture and Design	Certified – 6 Star Green Star, 2014	17,673 m^2	14,528 m^2	7

Table 1.
Description of case study buildings.

Green Star Performance rating tools [35]. However, BOSSA has wider applications, BUS surveys were used in this study due to the narrower scope, more-wider presence globally and acceptance of the respective survey, and the license rights available with or for the case study buildings [28, 34]. Apart from the BUS itself, an online survey deployed through Survey Monkey (cloud-based interface) was used by one of the case study buildings. Although the survey used was different for one building, their project managers allowed the researcher to add or modify survey questions as required before they were finally deployed. Hence, to ensure comparative study across all the buildings, their questionnaire was modified, and questions were added or removed to match the themes of the BUS format. So, four buildings used BUS and one building used an additional online survey as well as the standard BUS. BUS is available as hard copy survey and online; the buildings used the online survey rather than hard copy.

The surveys were used to evaluate the feedback of case study buildings, focusing on staff rather than staff and students. This was because staff spend more hours in a building either for teaching, research, or undertaking other professional activities. The BUS surveys used a 5-point Likert scale for measurement with 1 rating the highest dissatisfaction to 5 being the highest level of satisfaction. Statistical analysis of the occupant satisfaction survey data was carried out using SPSS Statistics (version 26) analysing via one-way ANOVA. Many studies [36–38] have proved that the Likert scale data can be analysed using parametric tests, as the analysis can vary based on how the responses are formulated. For example, in this research, the distance between each item category is considered "equivalent," and when a Likert scale is assumed to present symmetry of variables and is equidistant, it behaves more like an interval-level measurement in practice, being suitable for parametric tests [37, 38]. In this study, the distance between each item category is "equivalent" and the Likert scale was used to run parametric tests on the data collected. The other key primary criteria were to assess if the data were normally distributed within all groups. Hence, the Descriptive and Levine's Test for Homogeneity of Variances was carried out for all variables [36, 37]. The output displayed normal distribution using the mean values as the measure of central tendency, that is, the probability distribution. Correspondingly, as the data represented a Gaussian population (normally distributed), one-way ANOVA was used to determine the statistical significance of data [36, 37].

In addition to the surveys, qualitative tools in the form of walk-in discussions, stakeholder interviews with the building facilities department and design stakeholders, and focus group discussions were undertaken. The walk-in discussions assessed occupant's (including building's staff and students) responses in a more informal

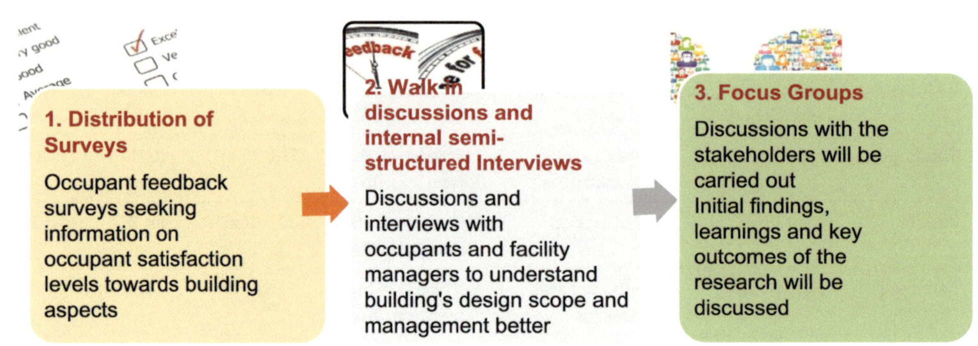

Figure 1.
Data collection approach.

manner, followed by the outcomes from the stakeholder interviews, assessing the role and responses of management and finally the focus group results. The focus group discussions were conducted to share the preliminary results and discuss preliminary results of the actual versus expected outcomes with key management stakeholders. These qualitative results assist in understanding and validating the occupant survey findings and facility management perspectives concerning sustainable and green building concepts. The study hence adopted an "Action Research" framework for data analysis. The steps involved in the data collection are summarized in **Figure 1**.

4. Results and discussion

The categories and sub-categories of the Green-Star As Design (v1) rating tool were used. Each sub-category has certain scores available that when combined form the maximum score available for that Green-Star category. The specific scores attained and statements made by the study buildings in the original Green Star application documents assist in determining the original design intention by a building.

Green-Star performance of case study buildings was analysed by utilising the Green-Star applications submitted by the respective institutions as an expression of interest for creating these buildings. The scores achieved by each building assists in understanding the commitments of the respective institutions. The mixed-method techniques deployed assisted in evaluating and validating how well these case study buildings performed in relation to the targets they laid out in their respective applications. The scores achieved by each case study building against the nine categories of the Green-Star As Design (v1) rating system are shown in **Table 2**.

As shown in **Table 2**, all the case study buildings scored high under the first five categories, namely – (i) Management – achieving score on factors, such as external and internal building commissioning, delivering appropriate management practices, evaluating user satisfaction, technical and physical performance; (ii) IEQ – for improved technologies for providing better IEQ services; (iii) Energy – achieving

Green Star As Design v1 Categories	Available Points	Points scored by Building A	Points scored by Building B	Points scored by Building C	Points scored by Building D
Management	14	12	12	13	14
IEQ	25	13	11	16	17
Energy	29	14	14	18	14
Transport	13	12	12	13	12
Water	13	10	10	16	14
Materials	20	12	12	14	14
Land use and ecology	7	4	4	2	4
Emissions	13	9	9	6	8
Innovation	5	0	0	0	10
Total	140	65 – 5 Star	63 – 5 Star	76 – 6 Star	83 – 6 Star

Table 2.
Scores achieved by each case study building across the nine Green-Star As Design (v1) categories.

Building Name	Total no. of staff	No. of responses
Building A	54	37
Building B	42	30
Building C	60	42
Building D	131	60

Table 3.
Case study building sample size.

energy efficiency, carbon footprint reduction; (iv) Transport; and (v) Water – using water-efficient technologies. The buildings scored moderate on materials, land use, ecology, and emission and no score on innovation, except Building D, which attained double the maximum score available for innovation, due to commitment in design and technologies promised.

The BUS surveys used to assess the staff responses towards these newly built Green Star rated buildings evaluated user comfort, indoor environment quality, wellbeing and productivity, and communication with management. In relation to the surveys, the effective population size (N) for each case study building is described in **Table 3**.

The key findings to emerge from the surveys were occupant responses towards three main themes – (i) impact of various design and work environment conditions (office/desk design, thermal comfort, acoustics, personal control, etc.) on their overall behaviour; (ii) their overall comfort and perceived productivity; and (iii) dissatisfied engagement levels from the management level

The user's perspective towards the impact of various design and work environment conditions on overall behaviour showed that 69% of users in Building A, 66% in Building B, 55% in Building C, and 58% in Building D rated the conditions in the building affected their behaviour personally (with colleagues) and professionally (in terms of work and productivity). Most users commented that the overall comfort or productivity was drastically affected, leading to changes in the work schedule and activities. Using Pearson (r) statistical test, significant at $p < 0.01$, the highest correlation was the variable 'building meets user needs and comfort', followed by 'overall comfort and perceived productivity', and 'perceived productivity and perceived health. The correlation results are presented in **Table 4**.

On rating the engagement of management, the mean of responses on the satisfaction scale of 1–5 for Building D for two variables were – 4.25 for: (i) facilities meet user needs, and (ii) speed of response for staff complaints, and 3.24 for the effectiveness of response. The results for Building D were the highest when compared to other

Variables	Correlation between variables			
	Building A	Building B	Building C	Building D
Building meets user needs and comfort	0.821	0.786	0.705	0.993
Overall comfort and perceived productivity	-	0.904	0.775	-
Perceived productivity and perceived health	0.913	0.747	0.863	0.797

Table 4.
Pearson correlation coefficient (r) between work environment variables in case study buildings.

Variables	Correlation between variables	
	Building B	Building D
Facility meeting user needs and effectiveness	0.955	0.738
Facilities meeting user needs with speed of response	0.929	0.681

Table 5.
Pearson correlation coefficient (r) between stakeholder engagement focused variables in case study buildings.

buildings, stating the highest level of satisfaction for the occupants of Building D. The other buildings also had mean scores close to or more than 3 on facilities meeting user needs and effectiveness of the response, demonstrating moderate levels of satisfaction. The occupants for all buildings rated lowest on the speed of response to a request to change, demonstrating lower levels of satisfaction. This particular aspect was also observed in the open-ended comments, where the occupants (average of 42% in study buildings) expressed their discontent towards the attitude of the facilities for not acting upon their requests or complaints on time. When the variables under this category were compared, a high correlation between two variables in each building was observed. The highest correlation (significant at $p < 0.01$) was observed in Building B followed by Building D demonstrating a high positive relationship. This suggests that for the users in Buildings B and D, the positive response for one variable affected the other directly. The correlation results are presented in **Table 5**.

On completion of occupant feedback survey analysis, it was observed that the occupants of the study buildings had mixed feelings about the facility. As primary occupants of the buildings, the staff were reasonably satisfied with the overall building, but not fully satisfied that it met their expectations specifically in relation to their office space and comfort. However, as per the occupant's perspective, the facilities were designed very efficiently and effectively for its students, with management's key focus being on providing a comfortable and desirable environment for the students over the staff.

Nonetheless, for all four buildings, the occupants on average showed high satisfaction (87%) towards the building design and Green Star attainment, medium/neutral satisfaction (57%) towards maintained temperature conditions, lighting levels, and comfort, and low satisfaction (42%) towards noise levels, office space, perceived productivity and health, and personal control over physical factors. This level of lower rating can be associated with several factors, including a shift to open-plan offices, sustainability initiatives not working well, and a lack of appropriate occupant engagement by the management, as per the occupant's open-ended comments provided in the surveys. Upon comparison between the four buildings, it was observed that Building D outperformed all other study buildings with the highest occupant satisfaction (89%) and Building C rating the lowest levels of occupant satisfaction towards most measured variables (54%).

Summarizing the responses, the quantitative results assisted in understanding the extent to which the study buildings met their occupant's needs and expectations, and how well have the facility managers succeeded in providing the building services to the occupants. On an average (43%), the results implied an overall low satisfaction of occupants with respect to their workspace environment and factors, demonstrating that the study buildings lack the ability to meet the primary objective of fulfilling occupant needs and improving their comfort and productivity. These quantitative results were triangulated with the qualitative results to eliminate any bias and understand the occupant-management relationships, and management's perspective in the study buildings for arriving at the final recommendations and conclusions.

The qualitative results obtained via walk-in discussions, stakeholder interviews, and focus group discussions re-emphasized the "missing link" or lack of communication loop between building occupants and the management, ultimately affecting the overall building performance. The management's priority is to achieve the University targets, whereas the occupants want a good indoor space to work, in addition to being able to express their needs and wants. Both managers and occupants' perspectives are individually valid and justified but do not complement due to low or negligible stakeholder prioritization and engagement in the management strategies. Hence, this chapter highlights the learnings from the case study buildings and develops a model using the best practices for improved execution in future projects.

The stakeholder interviews and focus group further assisted in triangulating the study's outcomes in correspondence to Green Star aspiration results. The results demonstrated that buildings could achieve certification, but the operation can still be poor in terms of low occupant satisfaction levels towards their workplace productivity and comfort. Major concerns observed in the walk-in discussions and stakeholder interviews were the lack of occupant engagement and consultations by the facility management officials during the building design and construction phase; none to extremely low involvement of design officials after building in use; and the limited focus on short term benefits over life cycle performance. The discussions with facility and building managers reiterated the key aspiration for the building's senior management was to achieve overall energy efficiency and make student-friendly facilities, rather than creating a productive workplace for the staff. The results also demonstrated that although the buildings scored high against the "management" and "energy" categories of the Green Star applications, most sub-categories were not satisfied in the operation phase. The management and design stakeholders primarily associated this non-functionality with technical defaults or cost-related issues.

The analysis reported in this chapter demonstrates that the development of the case study buildings has some success from an environmental, sustainability, and financial perspective. However, lacked drastically from an occupant (social) perspective. There are lessons that can be drawn upon for future developments to improve outcomes further. The analysis of survey responses and interviews highlights that it is crucial to close the loop between the performance measurement and performance management of buildings to achieve sustainability. Moreover, it is recommended to integrate stakeholder engagement and management at each phase of any project, to optimize the potential for all the stakeholders, and create an output beneficial and satisfactory for all.

The best practice model hence recommended by the study serves as a guidance document, providing structure and direction for the design professionals, and facility and building managers to prepare and implement their actions strategically, to achieve the desired performance of their buildings along with building occupant satisfaction. The model is represented in **Figure 2**.

Figure 2 demonstrates the structure of the key factors of the "best practice model" recommended by the study, based on the gaps identified earlier on and learnings from the study findings. It represents a continuing sequence along with the direction flow of the stages, tasks, or events in a circular flow, with each stage having the same level of importance, to ultimately achieve project success and performance management throughout the life cycle. This model has been derived based on the analysis of green-rated tertiary education buildings; however, it can be adopted by all commercial building types as it informs a general structural change required in an organization's

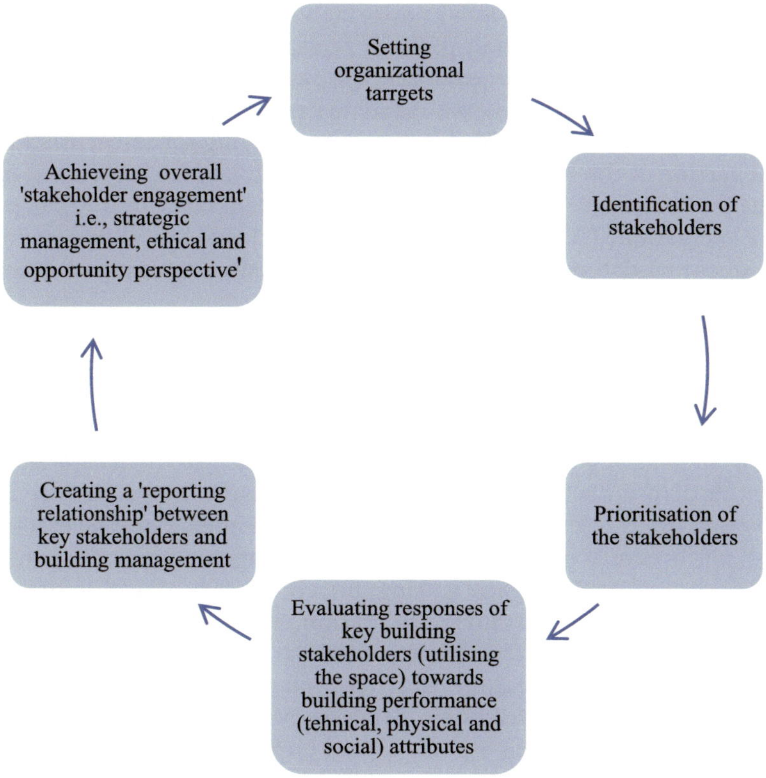

Figure 2.
Recommended 'Best Practice Model' for achieving sustainable building outcomes. Source: Authors.

management practice, to deliver required and appropriate governance strategies achieving all business outcomes and creating productive workplaces for the users. The key recommendations of the model can be summarized as a model that includes:

1. Identification and prioritization of key building stakeholders

2. Mapping the effects of building performance on different stakeholder groups and measuring the satisfaction of occupants (annually) using and being affected the most by building operation and function

3. Closing the loop of communication between the design team (internal and external) and building users by raising information

4. Collaboration across different organizations and agencies. This is because organizations or agencies often collect and manage their own data/research and have little or no incentive to share it with others. Without a culture of collaboration and knowledge sharing, attempts to understand and analyze interrelated building performance and sustainability issues are unlikely to succeed

5. A strong government mandate and institutional frameworks can provide a much-needed push

6. Routine (half-yearly or annually) performance monitoring of buildings as per the national guidelines for non-green or rating systems if certified to become mandatory for buildings. For example, the case study buildings were Green-Star rated (As Design) which does not mandate the buildings to evaluate performance once the certification is received, even when applying for re-certification. The recent version of the standard has been updated to mandate such assessments but the buildings previously certified are exempted, creating a gap in performance evaluation. This is a gap that needs to be closed by making evaluation techniques mandatory for buildings as per the desired or required standards

7. Appropriate incentive structure and practical arrangements for handling the information generated so that it becomes attractive and feasible to develop a set of good practices

Retaining and sharing the knowledge and lessons learned, then capitalizing and converting this to insights for future projects will be challenging, considering the flux in many organizations in these increasingly uncertain times. Nonetheless, incorporating ongoing performance evaluation, stakeholder engagement practices and understanding the need for reporting and getting the reporting relationship right in any organization becomes an important consideration.

The framework defined in this chapter is primarily based on the stakeholder engagement principles and is a three-step pyramid process. At the lower level or the base, the framework determines "Strategic Management" perspective, which simply means identifying organisational targets and priorities, enhancing decision-making processes, and ensuring stakeholder identification, prioritization, and engagement towards design, processes, and outcomes, followed by "Ethical" perspective illustrating maintaining harmony between stakeholder groups and identifying socio-behavioural aspects, and the higher level or apex constituting the "Opportunity" perspective illustrating knowledge sharing. The contribution of the research (based on the framework) to the knowledge will be the development of a best practice model for the built environment sector (applied/inspired from the academic sector) using a "holistic" or "integrated design" approach. Hence, the paper intends to recommend that being energy conservers and managing financial sustainability, the organization still needs to focus on engaging with all the stakeholders constructively.

Finally, the paper discusses the strategies to ensure replication of the respective best practice model to develop a supporting policy framework and build up capacity of building managers. The outputs generated by the paper are consistent with and instrumental to the achievement of the objectives that contribute to creating enabling environments for integrated sustainability strategies into building design and management.

5. Conclusions

The chapter presents an analysis of the gap between actual and expected performance in four Green Star institutions in Melbourne, Australia, and how POE generated results assisted in achieving this objective. Outcomes include the development of clear assessment mechanisms for establishing links between performance measurement and performance management with an understanding of how occupants view its value and description of a better performing building as compared to the others. The comparison was carried out on how well the building has been managed in conjunction with the sustainable building targets.

As observed by analysis of POE results of the case study buildings, it has been clearly indicated that the building users (mainly academic and non-academic staff) are not satisfied, and their needs have not been considered in the initial design brief. However, after triangulating the outcomes and studying the broader context, all four buildings have met their key parameters in terms of Green Star certification utility targets. What worked well for all buildings have been exceptional teaching and learning spaces, student study areas and institution image. Building variables that did not work well were lack of consultation with end-users and basic design faults or technical issues. This signifies that it is important to realize that the organizational facilities function cannot exist in isolation if the organization is to effectively exploit its entire asset base to best support the delivery of core services. Making constructive use of performance measurement results is critical if the organization is to improve the performance of its assets. At both individual and organisational levels, a better understanding of the relationship between what is being done and how well the organization succeeds needs to be developed. Based on the analysis, senior management and facility managers are seen to have a direct impact on building operations and performance. They can also adjust or redirect their strategies or identify new strategies for the organization.

Following the lead by the European Union since early 2021, there is a global momentum that is currently shifting towards Nearly Zero Energy Buildings, that is, high performing buildings utilising nearly zero or extremely low amounts of energy being met by renewable sources. While new commercial buildings in the US and Canada are now required to follow these principles, Australia is taking initiatives in the residential sector as well to achieve the same. Significantly, understanding the applications and potential of the building sector to effectively achieve overall sustainable outcomes. The outcomes of this study highlight that focusing on issues of building sustainability are not sufficient for a building's success. Being "green" is only one important feature of building success, but other aspects of building performance (user needs and engagement) must be considered as well. The focus on the green or sustainability aspect has detracted the construction industry from other equally important design issues relative to overall building performance. Understanding and taking the triple bottom line approach of being economically viable, environmentally responsible, and socially inclusive is important rather than simply aiming to be rated as "green" or sustainable. Furthermore, being sustainable and green should be translated to high-performance buildings for optimal benefits. Supporting this argument, the outcomes of this study are noteworthy for individual buildings be they academic or otherwise, and benefits wider government, non-government and market audiences seeking new knowledge about performance development.

Zero-energy programs often build on these frameworks supporting a variety of non-energy benefits relative to standard buildings, such as improved comfort, improved occupant health and productivity, enhanced indoor environment quality, and higher occupancy rates. Therefore, this study concludes that in making building codes, policies and management approaches more stringent, prioritisation and engagement of building stakeholders, and increased interest in decarbonisation, would assist in encouraging and assisting buildings that are truly zero-energy and zero-energy-ready.

Acronyms and abbreviations

ANOVA	Analysis of Variance
BREAM	Building Research Establishment Environmental Assessment Methodology
BOSSA	Building Occupants Survey System of Australia
BUS	Building User Survey
CASBEE	Comprehensive Assessment System for Built Environment Efficiency
COP	Conference of Parties
GBCA	Green Building Council of Australia
GHG	Greenhouse Gas
HK BEAM	Hong Kong Building Environmental Assessment Method
HQE	Haute Qualité Environnementale
IEA	International Energy Agency
IEQ	Indoor Environmental Quality
LCA	Life Cycle Assessment
NABERS	National Australian Building and Energy Rating Scheme
LEED	Leadership in Energy and Environmental Design
NZEB	Nearly Zero-Energy Buildings
POE	Post Occupancy Evaluation
SBAT	Sustainable Building Assessment Tool
SPSS	Statistical Package for the Social Sciences
UNEP	United Nations Environment Programme

Author details

Karishma Kashyap, Usha Iyer-Raniga* and Mary Myla Andamon
RMIT University, Melbourne, Australia

*Address all correspondence to: usha.iyer-raniga@rmit.edu.au

References

[1] UNEP & IEA. Towards a zero-emission, efficient, and resilient buildings and construction sector. Global Status Report. 2017. Available from: https://www.worldgbc.org/sites/default/files/UNEP188_GABC_en %28web%29.pdf [Accessed: July 23, 2018]

[2] GABC. Towards zero-emission efficient and resilient buildings. Global Status Report 2016. Available from: https://globalabc.org/resources/publications/gabc-global-status-report-2016-towards-zero-emission-efficient-and-resilient [Accessed: July 23, 2018]

[3] WGBC. About Green Building. The Benefits of Green Buildings. 2018. Available from: https://www.worldgbc.org/benefits-green-buildings [Accessed: April 27, 2019]

[4] IPCC. Summary for Policymakers: IPCC Special Report on Renewable Energy Sources and Climate Change Mitigation. 2011. Available from: https://www.ipcc.ch/pdf/special-reports/srren/SRREN_FD_SPM_final.pdf

[5] Akadiri PO, Chinyio EA, Olomolaiye PO. Design of a sustainable building: A conceptual framework for implementing sustainability in the building sector. Buildings. 2012;**2**:126-152

[6] Perera F. Pollution from fossil-fuel combustion is the leading environmental threat to global pediatric health and equity: Solutions exist. International Journal of Environmental Research and Public Health. 2017;**15**(1):16

[7] ICC. Overview of the IgCC. International Code Council: Washington, DC; 2018

[8] Munaro MR, Tavares SF, Braganca L. Towards circular and more sustainable buildings: A systematic literature review on the circular economy in the built environment. Journal of Cleaner Production. 2020;**260**:121-134

[9] Cheng B, Li J, Tam VWY, Yang M, Chen D, Bim A. LCA approach for estimating the greenhouse gas emissions of large-scale public buildings: A case study. Sustainability. 2020;**12**(2):685

[10] Yu S, Liu Y, Wang D, Bahaj AS, Wu Y, Liu J. Review of thermal and environmental performance of prefabricated buildings: Implications to emission reductions in China. Renewable and Sustainable Energy Reviews. 2021:**137**(C). DOI: 10.1016/j.rser.2020.110472

[11] WGBC. Building Ambition: The Future of the Built Environment Post COP26. 2021. Available from: https://www.worldgbc.org/news-media/building-ambition-future-built-environment-post-cop26

[12] Jenson, M. 5 Reasons Why Building Design is More Important Than Ever. 2014. Available from: http://sefaira.com/resources/5-reasons-why-building-design-is-more-important-than-ever/ [Accessed: May 15, 2018]

[13] DEE. Australia's 2030 Climate Change Target. Canberra, Australia: Department of Environment and Energy; 2015

[14] Abbaszadeh SL, Zagreus LD, Huizenga C. Occupant satisfaction with indoor environmental quality in green buildings. In: Proceedings, Healthy Buildings, Lisbon, Portugal. June 2006;**III**:365-370. http://escholarship.org/uc/item/9rf7p4bs

[15] Brown Z, Cole RJ. Influence of occupants' knowledge on comfort

expectations and behaviour. Building Research & Information. 2009;**37**: 227-245

[16] Mallawaarachchi BH, De Silva ML, Rameezdeen R. Importance of occupants' expectations for acceptance of green buildings: A literature review. In: The Second World Construction Symposium 2013: Socio-Economic Sustainability in Construction. Colombo, Sri Lanka; 2013

[17] ARUP. The User Factor: Effect of Buildings on User Performance. ARUP Global Research. 2016. Available from: https://research.arup.com/projects/the-user-factor-effect-of-buildings-on-user-performance/-overview [Accessed: October 4, 2018]

[18] Hansika WAM, Amarathunga PABH. Impact of Office Design on Employees' Productivity; a Case Study of Banking Organizations of North Western Province in Sri Lanka (December 8, 2016). University of Sri Jayewardenepura, Sri Lanka, 13th International Conference on Business Management (ICBM). 2016. Available at SSRN: https://ssrn.com/abstract=2910255

[19] Göçer Ö, Göçer K, Ergöz Karahan E, İlhan Oygür I. Exploring mobility & workplace choice in a flexible office through post-occupancy evaluation. Ergonomics. 2018;**61**:226-242

[20] Nooraei M, Littlewood JR, Evans NI. Feedback from occupants in 'as designed' low-carbon apartments, a case study in Swansea, UK. Energy Proceedia. 2013;**42**:446-455

[21] GBCA. Energy Consumption and Greenhouse Gas Emissions Calculator Guide. Australia: Green Building Council of Australia; 2016

[22] ACTS. ACTS Initiatives. QLD, Australia: Australasian Campuses Towards Sustainability; 2017

[23] Khoshbakht M, Gou Z, Xie X, He B, Darko A. Green building occupant satisfaction: Evidence from the Australian higher education sector. Sustainability. 2018;**10**:21

[24] Olanipekum AO. Motivating Project Owners to Increase their Commitment Towards Imporving the Delivery Performance of Green Building Projects. Brisbane, Australia: Doctor of Philosophy Queensland University of Technology; 2017

[25] Kajikawa Y, Inoue T, Goh TN. Analysis of building environment assessment frameworks and their implications for sustainability indicators. Sustainabililty Science. 2011;**6**:233-246

[26] NCC. Australian Government, National Construction Code; Green Building Council of Australia, Submission 99, p. 15. 2016. Available from: https://www.abcb.gov.au/ [Accessed: January 23, 2018]

[27] ASBEC. Low Carbon, High Performance – How building can make a major contribution to Australia's emissions and productivity goals. Australian Sustainable Built Environmental Council Report 2016. Available from: https://www.asbec.asn.au/research-items/low-carbon-high-performance-report/ [Accessed: July 23, 2018]

[28] Mitchell L. Green star and NABERS: Learning from the Australian experience with green building rating tools. Cities and Climate Change URS. 2009

[29] GBCA. What is green star. Green Building Council of Australia. 2019;**99**:3

[30] GBCA. Green Star Certification Process. Sydney, Australia: Green Building Council of Australia; 2009

[31] GBCA. Green Star Eligibility Criteria. Sydney, Australia: Green Building Council of Australia; 2009

[32] GBCA. The Value of Green Star – A Decade of Environmental Benefits. Sydney, Australia: Green Building Council of Australia (GBCA); 2013

[33] GBCA. Green Star Rating System. Sydney, Australia: Green Building Council of Australia; 2017

[34] BOSSA. What is BOSSA? An Australian Owned IEQ Assessment System. Australia: Building Occupants Survey System Australia; 2017

[35] Candido C, Kim J, de Dear R, Thomas L. BOSSA: A multidimensional post-occupancy evaluation tool. Building Research & Information. 2016;**44**:214-228

[36] Sivo SA, Saunders C, Chang Q, Jiang JJ. How Low Should You Go? Low Response Rates and the Validity of Inference in IS Questionnaire Research. Journal of the Association for Information Systems. 2006;**7**:351-414

[37] Carifio JP, Rocco J. Ten common misunderstandings, misconceptions, persistent myths and urban legends about likert scales and likert response formats and their antidotes. Journal of Social Sciences. 2007;**3**:106-116

[38] Ali N, Ahmed L, Rose S. Identifying predictors of students' perception of and engagement with assessment feedback. Active Learning in Higher Education. 2018;**19**(3):239-251. doi:10.1177/1469787417735609

Chapter 12

Energy-Efficient Retrofit Measures to Achieve Nearly Zero Energy Buildings

Nimish Biloria and Nastaran Abdollahzadeh

Abstract

Considering the 2021 IPCC report that justly attributes our deteriorating climatic condition to human doing, the need to develop nearly zero energy building (nZEB) practices is gaining urgency. However, rather than the typical focus on developing greenfield net-zero initiatives, retrofitting underperforming buildings could create significant scale climate positive impacts faster. The chapter accordingly discusses energy-efficient retrofitting methods under three categorical sectors—visual comfort (daylight-based zoning, shadings); thermal comfort and ventilation (solar radiation-based zoning, central atrium plus interior openings, insulation, and window replacement); energy consumption (efficient lighting system, and controllers, material and HVAC system optimization, PV panels as the renewable energy source). This chapter further substantiates these theoretical underpinnings with an implemented design scheme—an educational building within a cold semiarid climatic condition—to showcase the on-ground impact of these retrofitting strategies in reducing the energy used for heating and cooling and lighting purposes.

Keywords: architecture, energy retrofit, environmental quality, nZEB design methods, energy optimization

1. Introduction

The IPCC's sixth assessment report [1] has laid to rest any doubts about the genuine and very urgent impacts of climate change that our planet is facing. Global temperature change and changing precipitation patterns due to climate change have resulted in an increased frequency of extreme environmental events. Such devastating events have already taken up the form of large-scale forest and bush fires, hurricanes, increase in heat waves, droughts, water scarcity, and extreme storm conditions globally. According to the United Nations [2], climate and weather-related disasters have increased five-fold over the past 50 years. The impacts of this exponential increase have been clearly articulated by the World Meteorological Organization's Atlas of Morality: Economic losses from weather, climate, and water extremes (1970–2019)—between 1970 and 2019, natural disasters have accounted for 50% of global disasters (11,000 in total) resulting in 45% reported deaths (91% in developing countries) and 74% reported economic losses globally (amounting to $3.64 trillion) [3].

Interestingly, the WMO and the IPCC report have rightfully attributed the underlying cause for such widespread natural calamities to human doing. The WMO, after conducting a thorough review of the Bulletin of the American Meteorological Society concludes that within a time frame of 3 years (2015–2017) alone, a staggering 62 of the 77 disastrous natural events can be attributed to human influence. Sever heat waves impacts being the most apparent of these human impacts have even soared since 2015. The IPCC report further emphasizes the role of human doing attributing to global temperature rise and continued sea-level rise, which are seemingly irreversible over years to come. Human-made and natural wonders in the form of dense low-lying mega-cities, beautiful islands, delta regions as well as coastal regions are all now under-threat and are increasingly feeling the impact of climate change. The recent outcry from some such impacted countries for climate justice has been voiced in the recent COP 26 summit held in Glasgow.

Reducing current emissions by 45% is the new global challenge that must be attained to limit warming to the 1.5°C mark by 2100. This limitation is not surprising considering that most of the energy produced globally has been reliant on burning fossil fuels (primarily coal). A clear trend can be seen in the increase of atmospheric carbon dioxide concentration since the end of the eighteenth century and the beginning of the nineteenth century—this coincides with the time when coal came into everyday use [4]. The scientific principle behind the global warming trend is relatively simple to comprehend—burning fossil fuels, biofuels, and biomass release carbon that was otherwise sequestered, thus exponentially adding to the current stock of carbon globally. This burning process releases gases accompanied by tiny carbon particles (ranging from PM 10 to PM 2.5)—black carbon that tend to trap the sun's energy in the atmosphere (at a much higher rate than CO_2), resulting in an increase in temperature. Forest fires, transportation, industries, buildings, electricity, and heat production are all black carbon and greenhouse gas (GHG) sources.

Fossil fuels and the associated use of coal and petroleum play a vital role in contributing greenhouse gasses (GHG) and black carbon and are fundamental to be discussed in the context of this chapter. However, industries, such as agriculture, including animals, chemical-intensive farming, clearing forests for agricultural land, etc., are among the highest contributors of GHGs to the atmosphere. For instance, according to the United States Environmental Protection Agency (EPA), methane gas, produced during combustion processes and anaerobic decomposition, has 28–36 times more potential to result in warming than CO_2 [5]. Similarly, nitrous oxide, fluorinated gases, sulfur hexafluoride are all GHGs with higher potential to retain warmth in the atmosphere.

Within this context of climate change and the increasing responsibility on the shoulders of all citizens of planet earth, the building sector, coupled with the behavioral change we need to acquire toward the production, usage, and storage of energy, is of high importance for the future of our existence. According to the World Green Building Council's "Bringing Embodied Carbon Upfront" report [6], building and construction activities together account for 39% of energy-related CO_2 emissions while they use 36% of final global energy. Besides this, the building also accounts for approximately one-third of black carbon emissions [7]. Of particular, importance within this emission is the 72.5% share belonging to the residential sector owing to its propensity for energy consumption—the third-largest energy consumer sector in the world. According to the IPCC's report, energy consumption-related indirect emissions in residential buildings have quintupled while it has quadrupled for the commercial building sector (from 1970 to 2010). With a projected increase in the world population by half—almost 3.6 billion people toward the end of this century and a total of 11.6 billion people by the year 2100, the demand for housing and thus the increase in energy consumption and emission production could lead to disastrous climatic scenarios. Upfront carbon or the carbon emission released before the built asset/building is in-use is projected to constitute half of the carbon footprint of new constructions until 2050. The grave responsibility on the building sector thus revolves around addressing the tension between dwindling fossil fuel reserves and the ever-increasing energy demand. The Global Status Report 2017 published by UN Environment and the International Agency [8] further proposes that to address this complexity, the energy intensity per square meter of the global building sector needs to be improved by a minimum of 30% by 2030. Efforts to decarbonize the building sector or, in other words attaining a net-zero or nearly-zero building target is thus quintessential.

The chapter adheres to a hybrid methodology involving empirical research and simulation-driven design. A systematic review and data extraction from scientific journals, environmental organization websites, governmental regulatory bodies, and informational data specifically meant for the builder community are interfaced with a simulation-driven design for retrofitting an existing building. This interface was established to test theoretical and professional advice rendered through the study of literature and the actual impact of these propositions in the retrofitting of a common existing building typology. Section 2 of the chapter firstly reasons and establishes the need for retrofitting existing buildings. Once established, Section 3 systematically describes the concept of nearly-zero and situates the building industry within it. Section 3 emphasizes three essential components that need consideration while retrofitting existing buildings: Visual comfort (daylight-based zoning, shadings); thermal comfort and ventilation (Solar radiation-based zoning, openings, insulation, and window replacement); energy consumption (efficient lighting system and controllers, building material and HVAC system optimization, PV panels as the renewable energy sources). Section 3, while elaborating upon the technicalities involved in the three components, simultaneously presents the findings of the simulation-driven design as a proof of concept, thus elaborating upon the effectiveness of the promoted solutions. Section 4 serves as the conclusion of the chapter and future suggestions for developing a nearly-zero building future.

2. Retrofitting existing buildings vs. building new buildings

Given the context of climate emergency and the pivotal role that the building industry must play in reducing carbon emissions, the question is whether to focus on retrofitting existing buildings or tearing down existing buildings and building new net-zero buildings? Retrofitting existing buildings could offer an excellent opportunity for reducing our carbon emissions at a faster pace. The low rate—1% per year of building demolition and rebuilding in most parts of the world [9] and the fact that almost two-thirds of the existing global building stock will still constitute buildings that exist today are reasons enough to opt for the retrofitting path. Not attending to this significant building stock would imply continual carbon emission even in 2040, resulting in a failure in achieving the 1.5°C target set forth via the Paris agreement. The 2021 Pritzker Architecture prize laureates Anne Lacaton and Jean

Phillipe Vassal echo this view of retrofitting via their unique approach to building a smarter, greener, and inclusive built environment—never demolish, remove or replace, always add, transform, and re-use [10]. Countries like Australia, with almost eight million homes constructed from the 1950s onwards (like buildings elsewhere in the developed and the developing world), were built during a time with far less stringent regulations around energy and building standards, insulation, and material quality. This resulted in buildings either on the verge of collapse, being poorly designed, containing hazardous materials (such as asbestos), or consuming high amounts of energy to maintain comfort levels for their inhabitants. In the Australian context alone, buildings are associated with 18% GHG emissions and 20% final energy use (COAG Council report) [11]. According to an editorial published in The Guardian, the United Kingdom's trend to demolish 50,000 buildings per year to construct new ones is responsible for two-thirds of the total waste production of the entire country. The construction of new buildings is additionally associated with 10% of the UK's carbon emissions [12]. In the United States, almost seven million buildings are estimated to undertake remodeling and renovations in addition to commercial buildings undertaking capital improvements [13]. Such large-scale building alterations offer an excellent opportunity to include energy performance enhancements while conducting renovations one architectural element at a time (wall, roof, windows, floors, etc.)—Opportunistic Retrofitting. It is projected that three measures—re-siding, window replacement, and re-roofing, can cut result in 25% more energy savings alone. Bloomberg presents an easy to comprehend comparison for how much impact this 25% reduction will entail. Suppose this 25% energy saving potential is harnessed by even 1% of the US's 83 million existing single-family homes. In that case, it will reduce carbon emissions by more than 1.6 million metric tons yearly, which is equivalent to removing 350,000 passenger cars from highways. Besides savings on energy bills worth $400 billion each year, this positive environmental impact provides compelling arguments to transition to a nearly zero carbon building practice.

Besides this, it is essential to note that refurbishing and restoring existing buildings result in saving the embodied carbon footprint of the material used while constructing these buildings. This saving results in negating costs for mining, manufacturing, shipping, etc., of new materials that would otherwise be used for new constructions. In the long run, retrofitting thus becomes cost-effective with respect to CO_2 rather than building new. However, what is also vital to consider is the positioning of the building sector within the bigger landscape of energy and climate change debates. Retrofitting on its own, though beneficial, would benefit immensely if it harnesses an energy upgrade involving the following: Incorporating improvements in the energy efficiency of building operations; embracing a shift from fossil fuel to electric or district heating that is backed by carbon-free renewable energy generation practices; generation of carbon-free renewable energy on-site.

Within the current context of popular media exploding with discussions around climate emergencies and the need to reduce our carbon footprint and associated emissions, awareness about harnessing renewable energy has strengthened within the general population. However, what does it truly mean to become carbon neutral or, for that matter, what does a zero carbon badge imply for the building sector? The next section of this chapter engages in a short discussion around the concept of zero carbon to base tools and techniques that can be instrumental for reaching a zero carbon or a net zero retrofitting strategy.

3. What does being nearly zero implies in the context of a building?

The discussion thus far identifies why we need to reduce emissions and why we need to become highly energy efficient, especially when it comes to the building industry. The fundamental goal here is to neutralize resource consumption by reducing energy needs and harnessing renewable resources for energy production. This approach will produce buildings that offset the total amount of energy used by the building annually with the amount of renewable energy that can be captured on-site or via renewable energy providers [14]. The concept of net zero buildings (NZEB) was first discussed internationally in 2008 [15] and has been refined over time by the International Energy Agency (IEA) with almost 20 nations globally via the Task 40 initiative. The European Union was similarly discussing the definitions through its EPBD initiative that finally resulted in coining the term nearly zero energy buildings (nZEB) [16–18]. EPBD's recast directive establishes that the nearly zero or significantly less amount of energy required during the operation of the building should be catered to via energy derived from renewable energy sources (on-site or generated nearby). In Europe, the application of the nZEB model has become a requirement since December 31, 2018, for all public buildings. This application has slowly percolated to all new buildings from 2020 onwards. The United States Department of Energy (DOE) further classified zero energy buildings based on their total life cycle energy. This definition included building energy (on-site building energy consumption—heating, cooling, ventilation, indoor and outdoor use, lights, plug loads, process energy, elevators, intra-building transportation, etc.), energy consumed in transportation of primary fuels, thermal and electric losses in generation plants, and loss of energy during transport of energy to the building site [19]. A holistic spin on energy balancing of the building is thus proposed. It is also important to note that as opposed to autonomous zero energy buildings that can generate and consume equal amounts of energy to sustain themselves, nZEBs can connect to the external electricity grid provided that the annual energy export is equal to the annual energy import.

Accordingly, this chapter focuses on the added value of an nZEB by retrofitting the existing stock of buildings by reducing their energy needs and employing appropriate physical improvements to enhance its efficiency standards. In retrofitting existing buildings, three fundamental principles need to be adhered to—reuse, reduce, and sequester. Re-use in this context implies the use of recycled materials, paying specific attention to the end-of-life re-use properties of the materials used during retrofitting, and the idea of designing with an aim for deconstructing. Reducing implies carefully optimizing materials used during the renovation to selectively opt for low carbon materials [20]. Sequestering in the case of retrofitting involves the provision of carbon sequestering locations coupled with materials that can sequester carbon, such as bioplastics, the use of mycelium insulation, recycled plastic, and biomaterials-based carpeting, and 3D printed wood made from sawdust, to name a few. Ideally, retrofitting to reach a nearly zero energy building status can be clubbed into two design strategies—passive and active [19]. Passive strategies incorporate material properties, urban positioning/orientation, envelope design, and shading, to name a few. On the other hand, active strategies deal with improvements within HVAC systems, energy-efficient lighting, etc.

On the material front, though, one needs to comprehend the notion of "Embodied Energy." Embodied energy is typically associated with the total impact of material greenhouse gas emissions during its entire life cycle. Lifecycle covers the dimensions of a material's extraction, manufacturing, transportation, construction, maintenance,

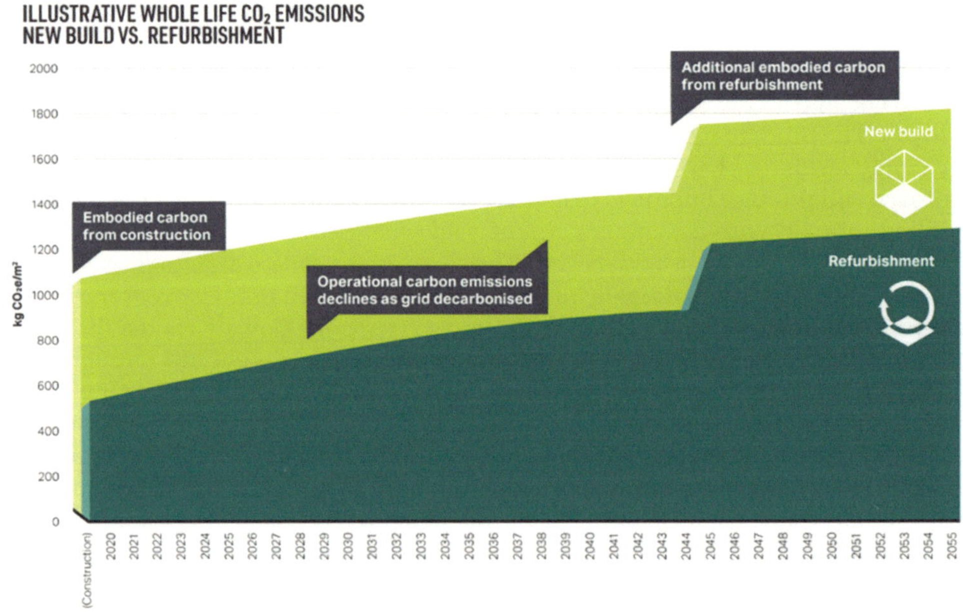

Figure 1.
CO_2 emissions of new build and refurbishment (image source: AECOM, 2021 [22]).

and disposal. If we take the new construction route, it is estimated that embodied carbon alone will be responsible for 72% of the carbon emissions between now and 2030 [21]. Embodied carbon cannot be rectified because it is embedded within the building once it has been erected. It is thus crucial to address the embodied carbon issue during the design or before the retrofitting stage is actualized on any building site (**Figure 1**).

Besides this, the basic building materials which are prevalently used in the construction sector—concrete, steel, and aluminum, are together responsible for 23% of total building emissions in themselves. Portland cement, the primary ingredient for making concrete, is responsible for releasing 40% CO_2 during the burning of fossil fuels for its manufacturing and emits 60% CO_2 during its processing phase. Similarly, the production of steel is a significant determinant of how much CO_2 it generates. Typically, basic oxygen furnaces responsible for producing steel rely on burning fossil fuels—coal or natural gas to melt iron ore, thus contributing to CO_2 emissions at a large scale. Better material alternatives or the alteration of production technologies of such fundamental materials used in the building industry are thus crucial. Embodied energy becomes specifically vital if seen from the context of developing nations, witnessing a boom in the building construction sector at an exponential rate.

Having gained some perspective on the concept of nearly zero energy buildings, the big question then is how do we translate this theoretical thinking into reality via retrofitting existing buildings? The following section provides some perspective on the same.

4. Retrofitting toward a nearly zero energy future

In retrofitting a building, one of the critical aspects to consider is the material properties of the existing building and the carbon content typically residing therein.

The London Energy Transformation Initiative (LETI) proposed an embodied carbon budget of 600kgCO_2e/m^2 for us to attain our carbon reduction goals [22]. To understand the feasibility of achieving this low figure, one must comprehend the total embodied energy typically present in an old building. Aecom, in a web article titled "The carbon and business case for choosing refurbishment over new build," breaks down the embodied carbon within various components of typical residential buildings: Frame 24%; substructure 19.6%; upper floors (14.9%); building services (13.4%); internal finishes (12.4%); external walls; windows and doors (8.8%); roof (5%); fitting and furnishings (1.2%); internal walls and doors (0.6%). The elements typically associated with the highest embodied carbon (substructure, frame, upper floor, and roof) are candidates that qualify for retrofitting to save on emissions produced during the breaking down of an old building and re-constructing these buildings elements while building anew.

To holistically retrofit existing buildings for enabling the transition toward a nearly zero energy building, three main scopes encompassing passive and active strategies should thus be considered:

a. Visual comfort (daylight-based zoning, shadings)

b. Thermal comfort and ventilation (solar radiation-based zoning, openings, insulation, and window replacement)

c. Energy consumption (efficient lighting system and controllers, building material and HVAC system optimization, PV panels as the renewable energy sources)

To explain these three strategies, the chapter, apart from providing theoretical advice, also elaborates upon the results of a simulated case study—an educational building in Iran. An initial comfort analysis in terms of visual and thermal condition and energy use of this existing building provides insights on the required improvements and thus informs retrofitting strategies. The elaborated project deploys the three-stage retrofit process on an educational building in Tehran (35.6892° N, 51.3890° E), Iran, in a cold semiarid climatic condition (Köppen climate classification: BSk). The process adopted by the authors incorporated radiance and energy plus simulation engines. According to the weather data for a typical year in this area, January is the coldest month, and August represents the hottest time of the year, each with an average monthly temperature of 3.89°C and 30.07°C. The site also enjoys a high level of solar exposure and experiences cloudy sky conditions only 15% of the time.

Situating a to-be retrofitted building within its environmental context is an ideal strategy for understanding the reasons behind its current energy performance. To strive toward an energy-efficient status for an existing building, both interior and exterior aspects are equally crucial to be considered. Visual and thermal comfort components can be primarily linked with the urban positioning of the building itself and imply conducting on-site solar radiation analysis for extracting the degree of solar exposure received by the building's external facades. In addition to this, mapping the demand for lighting and thermal energy of the interior programs of the building is vital. This step aids in making informed decisions to re-position or augment a building's program to take the best advantage of the building's solar exposure while naturally reducing the amount of energy required to heat or cool the building's interiors.

For instance, in our educational Building case study, solar radiation analysis (**Figure 2**) revealed that the Southern zones witnessing the highest solar radiation

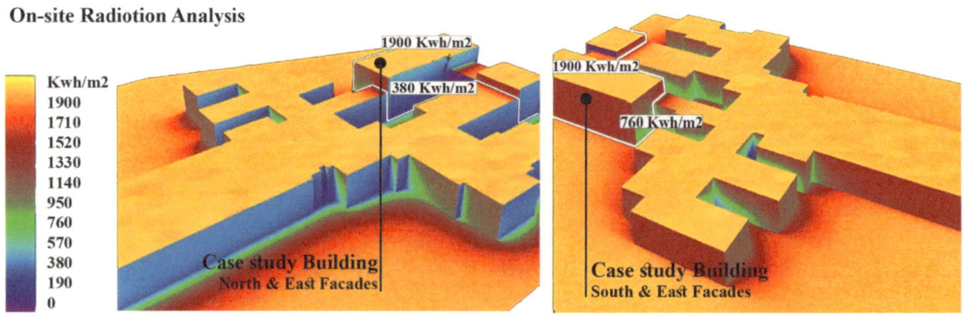

Figure 2.
Solar radiation analysis on building envelope.

constituted of primary programs and were induced to a high level of daylight and solar heat gain. On the contrary, northern zones were the coldest spaces with a pleasant daylight quality without experiencing visual discomfort caused by glare. The Eastern and Western zones were typically dedicated to circulation and service areas. Daylighting and thermal requirement-based zoning and categorization of the building's program are also conducted to determine the ideal spatial distribution of the program within the building (**Figure 3**).

The other important aspect of understanding visual comfort is calculating daylight quality to evaluate illuminance levels and glare probability and subsequently calculate thermal conditions to evaluate the total amount of discomfort hours within the primary programs of a building. It is vital to educate and to understand the importance of using the sun for solar tempering. Working with rather than against natural solar movement and exposure also aids in achieving energy savings otherwise required for heating purposes, and appropriately shading also aids in reducing cooling requirements. A strategic manner of avoiding added costs and energy expenditure for added thermal mass required for maximizing passive solar heating can thus be achieved. Such concerns are primarily best addressed during the design phase of the proposed retrofit.

In the case of the studied institution building, the south-facing spaces tend to receive excessive sunlight and suffer excessive visual discomfort (ASE > 10%, LEED 0 point). Therefore, the use of optimum shadings is suggested in the retrofit process. Despite this, the northern zone achieves all 3 points of the LEED rating system, which indicates that both sufficient daylight level (sDA > 75%) and visual comfort

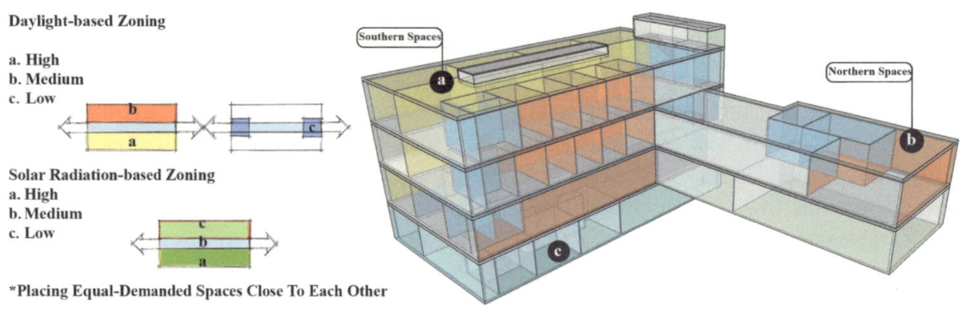

Figure 3.
Spatial zoning based on the solar geometry.

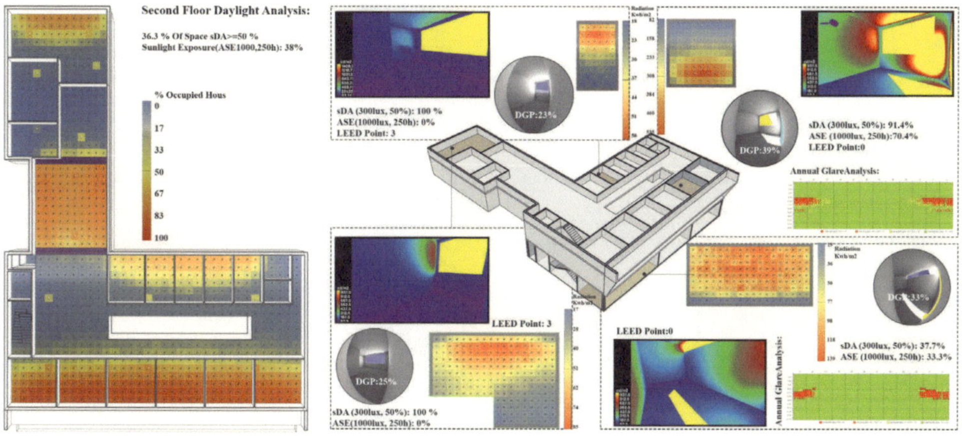

Figure 4.
Visual comfort (daylight distribution and glare) assessment.

(ASE < 10%) is achieved in these zones. Therefore, it is highly suggested to consider solar geometry in the renovation process to establish the feasibility of removing or adding the interior partitions. An overall analysis of daylight distribution on each floor (**Figure 4**) can elaborate on the optimal positioning of the interior program of a building in relation to solar radiation. Such strategic thinking and informed decision-making pertaining to program positioning before beginning the retrofitting process can thus ensure reducing the total energy consumption of a building considerably.

Thermal comfort analysis by assessing the average monthly temperature within an existing building's interior spaces can further help designers to understand overcooling or overheating scenarios throughout the year. Similarly, an annual analysis of discomfort hours also provides information on thermal conditions on a dynamic scale and can show critical thermal condition levels of interior space. Thermal imaging using an infrared camera can also be deployed for measuring on-site thermal conditions in individual rooms. Typically, such analysis can also be categorized under an energy audit of an existing building. For the case of the educational building understudy, the annual discomfort hours were calculated as 988.73 hours (**Figure 5**) and were associated with the average monthly temperature and the percentage of occupancy time that the hours corresponded with. The design phase of the retrofit being a calculated experimental phase also renders itself for making design decisions pertaining to elements such as atriums, window opening sizes, etc., while keeping in mind the window to wall ratio. Such additions, typically aimed at improving natural ventilation, can dramatically improve inside temperature conditions, thus impacting the total energy required for heating, cooling, and artificial ventilation. In the case of the educational building, the insertion of a central atrium combined with interior windows with a window to wall ratio of 20% was experimented with as a designed addition. A computational fluid dynamics (CFD) simulation analysis on a typical spring day (without an HVAC system operational) was able to provide a good overview of well-ventilated zones vs. zones which needed further improvement strategies such as supply vents etc., (**Figure 6**). Such initial analyses of the indoor environmental condition thus offer a basis for taking calculated decisions of the required retrofit strategies on a case-by-case basis.

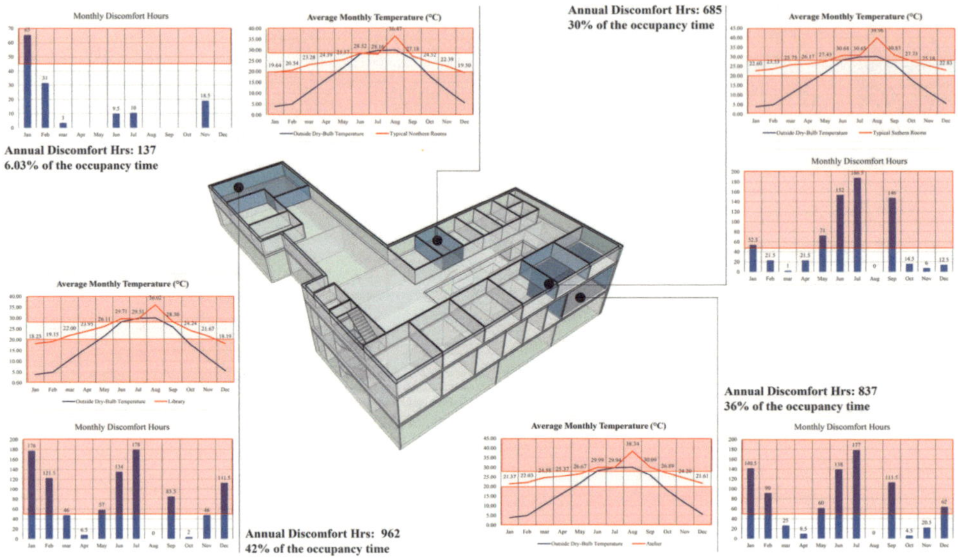

Figure 5.
Thermal comfort assessment.

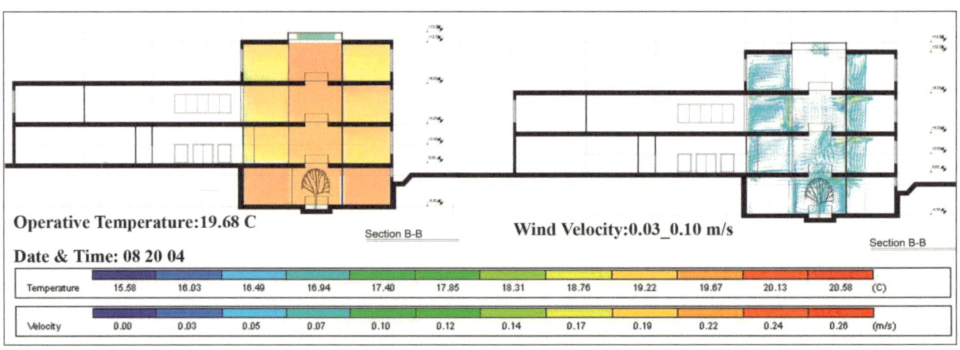

Figure 6.
Ventilation (velocity and temperature) assessment in the central atrium.

Tangible aspects connected with thermal comfort can be further categorized into lighting, sealing, insulation, and window replacement components. Lighting inside a building is equally crucial. Older buildings typically make use of fluorescent lighting and light bulbs. These tend to consume much more energy than contemporary energy-efficient CFL or LED light bulbs that last longer and are mercury-free. Retrofitting should thus ideally involve replacing existing fluorescent lighting systems with LEDs equipped with linear controllers. This simple change can almost halve the energy used for lighting purposes. Using motion sensors in areas where lights are left on often also allows energy to be saved. For the case of the educational building, this small change in lighting coupled with a daylight sensor for controlling the intensity of lighting in real time can halve the total energy consumed (**Figure 7**).

Sealing the building envelope is another step that is a highly efficient and cost-effective measure that can be deployed during the retrofitting phase for any building. Saving valuable energy required for heating and cooling and improving comfort,

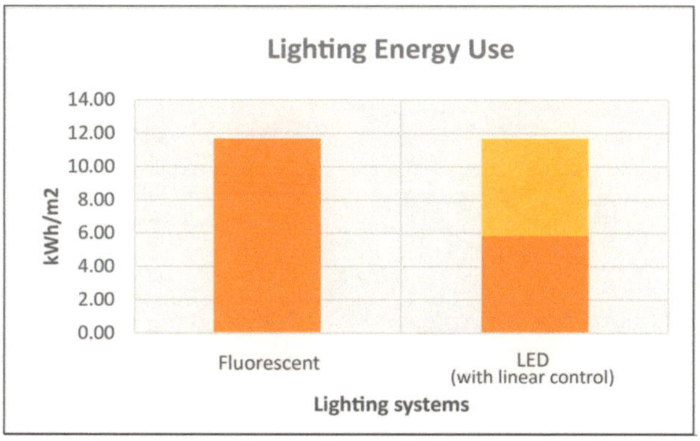

Artificial Lighting:
LED With Linear Control
Model: AOK(Factory) LED 2017
Life Time: 50000 (Standard;TM21 L70)
Luminous Distribution: 120◦
CRI: 7080/
Efficacy: 130 Lm/W
Power Density w/m2 100lux : 2.5
Daylight Sensor
Model: PIR 150W

Figure 7.
The energy use of reference (fluorescent) VS. optimized lighting system (LED with linear control).

reducing noise, and improving air quality are all direct impacts of sealing the building envelope, resulting in nearly zero energy targets. Often a blower door test can be conducted to evaluate the air leakage during air change per hour. Air change here implies the volume of air that equals the house volume exchanged with the outside air [23]. A blower door can establish a negative 50-pascal house pressure. Existing homes in need of retrofitting can often leak air at the rate of 15 air changes per hour (15 ACH50). Sealing the external and internal surfaces of a building can aid in bringing this level of leakage down between a range of 2ACH50 (airtightness standard for a cost-effective zero energy home)—0.7 ACH50 (airtightness standard for a passive house). Setting an airtightness goal for retrofitting projects can thus prove to be a wise decision for cutting down on energy consumption by optimizing the building envelope.

Building insulation should be considered as a fundamental component of any retrofit project. Different surfaces, such as walls, floors, and ceilings, require different types and thicknesses of insulating materials that are contextually derived based on climatic conditions and solar exposure. The R-value, or in other words, the ability of a material to resist the flow of heat, is important while choosing adequate insulation and is highly dependent on the kind of material used for insulation rather than the thickness of the material used. The climatic context of the region within which the retrofitting needs to be undertaken thus plays a significant role in determining the requisite R-value of the chosen insulation type and thickness. The Zero Energy Project report [24] outlines practical ways in which high-performance walls (exterior rigid insulation; single plate, double stud walls; double plate walls), high-insulated ceilings (blow insulation onto flat ceilings; insulating cathedral ceilings; exterior rigid insulation), and high insulated-floors (insulated slabs; insulated basements; crawl space), can aid in reducing energy loads by means of the application of appropriate insulation.

Similarly, window replacement and door replacements can play an essential role in transitioning to a nearly zero energy building. Organizations such as the National Fenestration Rating Council have contributed heavily toward establishing rating systems of window and door performance measurements in the form of labels affixed to off-the-shelf window and door units [25]. This process aids in the simplification of retrofitting wherein the everyday citizen can make informed decisions pertaining to the efficiency of these quintessential components of a building. Like the R-value of insulation, a U factor is of prime importance as it indicates the efficiency of a window

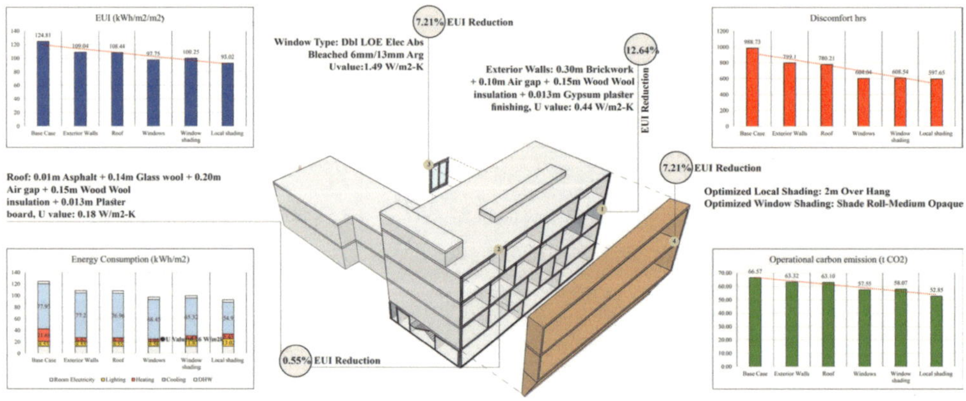

Figure 8.
Construction material optimization.

as regards heat escaping from the interior of a building. For visual comfort purposes, the visible transmittance value is also associated with window ratings. It is responsible for measuring the effectiveness of a window to light the interiors of space with daylight. For doors, the solar heat gain coefficient value that demarcates the door's resistance toward unwanted heat gain and the air leakage value that indicates the entry of external air through the door are vital measures specified by such councils.

For the case of the studied educational building, to establish a lower heat transfer threshold, the application of interior insulation (0.10 m air gap +0.15 m wood wool), replacement of windows with highly sealed windows with a low U value and coated with efficiency-enhancing coatings (Dbl LOE Elec Abs Bleached 6 mm/13 mm Arg), provision of interior window shading (shade roll-medium opaque) and exterior—localized shading elements (2 meters overhang), were explicitly deployed for enhancing the efficiency of south facing spaces (**Figure 8**).

The energy consumption component for a building retrofit process involves active strategies. HVAC systems are omnipresent in the majority of homes globally and how to reduce the energy required for heating or cooling purposes is of particular importance here. The strategies—visual and thermal comfort enhancements, already contribute to reducing conventional HVAC systems' load. However, apart from these passive measures, selecting appropriate HVAC systems conducive to the climatic context and the proportion of spaces to be conditioned are essential criteria to consider. For instance, for residential properties, air-source heat pumps are highly efficient and can take up the form of mini-split heat pumps for individual rooms or multi-zone installations. Variable speed operation by means of sensing temperature conditions inside a building and accordingly increasing or decreasing heating or cooling speeds results in air-source heat pumps in achieving energy savings.

Other strategies such as working with combinations of different heating and cooling systems per the degree of solar exposure and desired comfort levels could also be experimented with. For the case of the educational building, four different systems of radiator + evaporative cooling (the most common system used in the location), VAV, fan coil, and heat pump were simulated and optimized to establish the most efficient option. Accordingly, the optimum system of unitary heat pump can reduce heating, cooling, and the total energy use intensity (EUI) by 69.03%, 38.21%, and 28.81%, respectively. The final results indicated that the proposed method could reduce the annual energy consumption (EUI) by almost half while doubling the comfortable

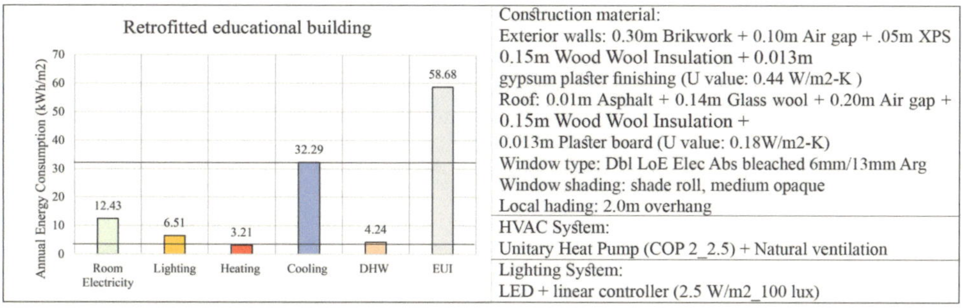

Figure 9.
Energy use and physical features (construction material, HVAC, and lighting systems) of the retrofitted case study.

hours indoors. **Figure 9** represents the energy consumed by different energy-consuming components of the retrofitted case study and the suggested replacement of physical features of the building. Such informed and analytically validated suggestions can inform property owners and make them aware of the retrofitting process's impact in transitioning to a nearly zero energy future.

Another active mode of energy generation is the renewable energy sector. Harnessing the sun's power by means of solar photovoltaic (PV) panels is one of the most cost-effective modes of harvesting renewable energy. The efficiency of solar panels is typically dependent on the amount of unobstructed solar radiation captured by the panels over a period throughout the day. After calculating the amount of energy conserved by applying the aforementioned passive and active strategies, a well-thought-out plan for solar energy capture must be developed. This is also due to the limitations of existing homes regarding the amount of open and exposed roof surface square meters available for the installation of PVs. A well-developed plan can aid in calculating the exact number of panels needed to manage and balance out the amount of energy required to reach a near-zero energy target. The inclusion of microinverters rather than centralized inverters should become the norm to encourage capturing optimal performance per panel while future-proofing the ability to add more panels in the future. Governments globally are now encouraging the installation of panels by providing subsidies and encouraging schemes that make solar leasing affordable and easily accessible.

For instance, in the case of the educational building understudy, PVs were suggested to make the most from renewable energy sources, such as solar radiation, to transfuse an annual amount of 10.7 mWh of electricity to the grid supply a part of the projects' total energy consumption. Hence, 24 panels are suggested to be placed on the roof to bring it closer to an nZEB design. Accordingly, the final design incurs less than 55 kWh/m^2 energy consumption, from which 9.32% is supplied by harvesting solar energy. The associated carbon emission (operational) was also reduced by 17.96% (**Figure 10**).

As a proof of concept for the propositions made in this chapter, jointly—both passive and active strategies proposed for the retrofitting of the educational building exhibited the potential to reduce the energy consumed for heating, cooling, and lighting purposes up to 85.19%, 58.57%, and 23.68%, respectively, compared to the base case (**Figure 11**).

Furthermore, the annual EUI could effectively be reduced by 52.98%, while the associated carbon emissions (t CO_2) and annual comfort hours also exhibit

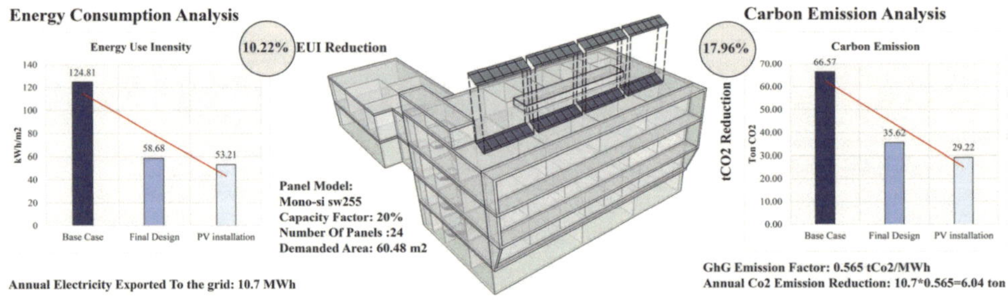

Figure 10.
Energy savings and CO_2 emissions reduction by PV installation.

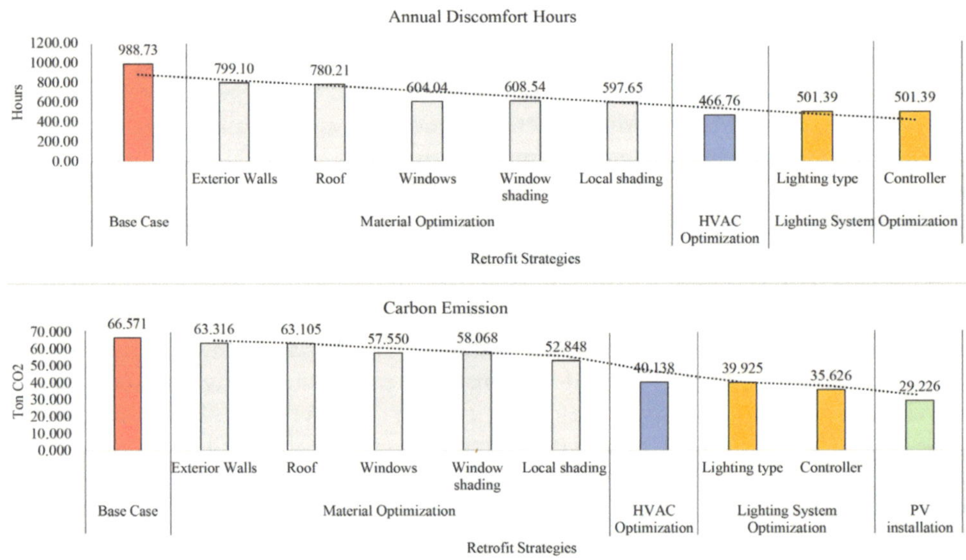

Figure 11.
The impact of the proposed retrofit strategies on the annual discomfort hours and CO_2 emissions.

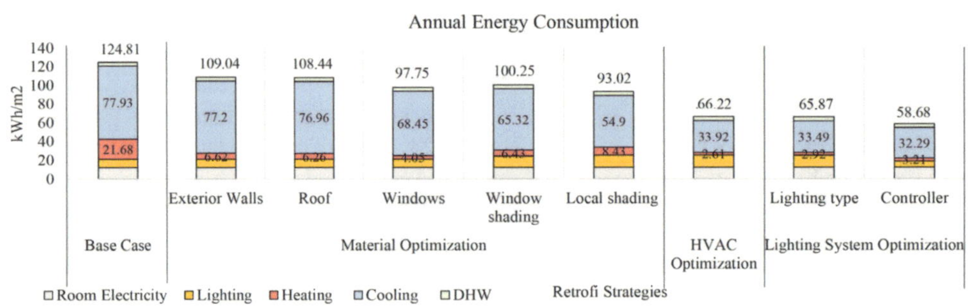

Figure 12.
The impact of retrofit strategies on annual energy use.

improvement by 46.48% and 49.28%. Replacing existing windows with highly efficient windows proved to have the highest impact (22.58%) on the comfort experienced indoors, followed by using an optimum HVAC system and applying interior insulations, each with an effective percentage of 21.90% and 19.18%. The inclusion of an efficient HVAC system also aided in reducing operational carbon emission to a great extent (24.05%). Installing PV panels also exhibited a substantial reduction in carbon emissions (19.18%) (**Figure 12**).

5. Conclusion

The chapter attempts to rationalize and strategize the retrofitting of existing buildings as a valuable means for transitioning toward a nearly zero energy future. The perspective presented in this chapter revolves around three fundamental components:

a. Visual comfort (daylight-based zoning, shadings)

b. Thermal comfort and ventilation (solar radiation-based zoning, openings, insulation, and window replacement)

c. Energy consumption (efficient lighting system and controllers, building material and HVAC system optimization, PV panels as the renewable energy sources)

An actual case study of an existing educational building has been conducted to present the tangible benefits of applying passive and active measures within these three components to reduce energy usage and carbon emissions. The results of this study have been presented in parallel to the theoretical discourse on methods to achieve a nearly zero energy building goal as a proof of concept for the advocated practices. Besides the case study conducted by the authors to reassure the readers about the benefits of retrofitting, Harvey (2013) further outlines various building typologies that have achieved energy reductions by adopting retrofitting practices—detached and single-family homes (50–75%); apartments (80–90% reduction in heating); building envelope retrofitting (1/2 to 1/3rd reduction in cooling, and 2/3rd reduction in cooling); HVAC optimization-based energy saving in commercial buildings (25–50%); lightning-based retrofitting in commercial buildings (30–60%) [9].

Retrofitting is a viable option to consider despite high upfront costs since the annual cost savings on the energy present an economically attractive scenario. However, critical mediation stages during the lifespan of a building must be identified since these can serve as potential stages to upgrade energy provisions. Such stage-wise upgradation can be streamlined to minimize disruptions to owners and organizations while keeping abreast of the latest technologies and techniques for energy conservation. Policy interventions that are participatory development-driven—between government, local councils, and owners/organizations, can further aid in contextually sensitive retrofitting processes. Bottom-up policy initiatives that subsidize and acknowledge geo-location, climatic and socioeconomic conditions, major renovation cycles, capital improvement cycles, and resiliency upgrades should undoubtedly become the norm in the near future. The strategies suggested in this chapter can further aid in systematically fusing passive and active approaches toward nearly zero energy buildings. Such strategies will also benefit substantially by interfacing them with qualitative research conducted on-ground that predominantly deals with the assessment of human behavior and the drive to adopt retrofitting strategies. Community concerns, economic limitations, fear of disruption of everyday life, etc., could become critical insights from such qualitative explorations. These can further aid in tailoring policies while being sensitive toward the concerns of the everyday citizen. Retrofitting processes can benefit building typologies, such as aging building stock (houses and apartments alike), large-scale institutional buildings, offices with older energy-intensive energy systems, and heritage buildings. Additionally, buildings located in zones that face severe weather conditions (extreme heat or cold) or are undergoing post-disaster reconstruction can also benefit through undertaking the nearly zero energy transition.

The need to address climate change to shape a sustainable present and a thriving future is of utmost importance now more than ever. Retrofitting existing building stock in conjunction with sensitizing citizens and corporations alike and the participatory development of building policies and programs could undoubtedly hold the key to reducing emissions. Let us never forget that we have only one Earth, and it is our collective responsibility to protect this beautiful planet and our future, which are intrinsically linked.

Author details

Nimish Biloria[1,2]* and Nastaran Abdollahzadeh[1]

1 Faculty of Design, Architecture and Building, University of Technology Sydney, Sydney, Australia

2 Faculty of Engineering and the Built Environment, University of Johannesburg, Johannesburg, South Africa

*Address all correspondence to: Nimish.biloria@uts.edu.au